THE ICE THAT BURNS

Investigating Rare Exothermic Reactions and Their Misunderstood Danger

Ingrid Solheimsdottir

This book is a work of nonfiction. The events, incidents, and scientific findings described herein are drawn from publicly available investigative reports, peer reviewed research, and documented historical record, as detailed in the Note on Sources. Every effort has been made to ensure the accuracy of the factual content presented.

This book is intended for general informational and educational purposes only and does not constitute professional, safety, engineering, or legal advice. Readers with specific safety concerns regarding chemical storage, handling, or industrial processes should consult qualified professionals and applicable regulatory guidance.

Author monogram and cover design by Ingrid Solheimsdottir.

Contents

Introduction

The Fire Hidden in Ice

On a January morning in 2007, a chemist named Robert would have told you, if you had asked him, that his reactor was under control. The gauges said so. The procedure said so. The batch had run one hundred and seventy four times before without incident, and there was no obvious reason why the one hundred and seventy fifth would be any different. Then the cooling system faltered, and a chemical reaction that had been quietly producing heat inside a sealed vessel found itself with nowhere to put that heat. Temperature climbed. Pressure climbed with it. Within minutes the reactor became a bomb built entirely out of its own contents, and when it let go, the blast was felt fifteen miles away in downtown Jacksonville. Four people died. Twenty eight more were hurt in buildings a quarter mile distant, struck by walls and windows that had no business being part of a chemical accident at all.

This is a book about that gap between what a reaction looks like and what a reaction is doing. It is a book about heat that does not announce itself, about processes that seem finished when they have only just begun, and about the particular danger of anything that gives off energy quietly, continuously, and for longer than anyone expects. The title borrows a paradox on purpose. Ice does not burn. Everyone knows that. But cold hay bales, damp grain in a steel bin, a jar of solvent pushed to the back of a supply cabinet fifteen years ago, all of these can sit at room temperature, feel utterly inert to the touch, and still be industriously generating the exact conditions required for fire. The heat is there before the flame is. That is the whole of the danger, and it is also the whole of the subject of this book.

Exothermic reactions are not exotic. That is the first thing worth saying plainly, because most people encounter the word only in a chemistry classroom, usually attached to a demonstration involving a beaker, some fizzing, and a teacher explaining that heat was released. The classroom version is accurate as far as it goes, but it also does something slightly misleading: it makes an exothermic reaction look like an event, something with a clear beginning and a clear end, something you can watch happen in ninety seconds and then write a lab report about. Most of the exothermic reactions that matter in the real world do not work that way. They are slow. They can take hours, days, weeks, or in the case of some industrial solvents, years. A rusting nail is an exothermic reaction. So is a compost heap breaking down kitchen scraps into soil. So is a barn full of hay quietly fermenting itself toward a temperature at which it can ignite without anyone striking a match. The energy release in all of these cases is real, measurable, and governed by the same thermodynamic principles that govern the reaction inside a rocket engine. The only difference is speed, and speed is exactly the variable that human intuition is worst at judging.

We are, as a species, extremely good at recognizing fast danger. A wildfire crowning through a forest canopy, a gas leak igniting, a firework misfiring, these register instantly as threats because they present themselves the way threats are supposed to present themselves: loudly, visibly, and with enough warning for the amygdala to do its job. We are considerably worse at recognizing slow danger, particularly when that danger is wrapped in something ordinary. A pile of oily rags in a garage does not look like anything. It looks like laundry that has not been dealt with yet. A bag of grass clippings left too long in a plastic bin does not look like a hazard, it looks like yard waste. A bottle of diethyl ether that has been sitting on a laboratory shelf since a researcher retired eleven years ago does not look dangerous either, it looks like an unglamorous bit of housekeeping that nobody has gotten around to. And yet each of these ordinary, unremarkable objects is capable, under the right conditions, of producing enough heat or enough shock sensitivity to seriously injure the person who eventually has to deal with them. The chemistry does not care that the object looks boring. The chemistry has been running the entire time.

This book exists to close that gap between appearance and reality, and it does so by moving through a series of specific, real, well documented cases. Every incident described in the chapters that follow actually happened. The people who died at T2 Laboratories in Jacksonville were real people with real names, and their deaths were investigated in detail by the United States Chemical Safety and Hazard Investigation Board, whose report forms the backbone of the chapter devoted to industrial runaway reactions. The explosion that leveled a fertilizer distribution facility in the small town of West, Texas, in April 2013, killing fifteen people including twelve emergency responders, is documented down to the estimated tonnage of ammonium nitrate involved and the composition of the wooden bins that failed to contain it. The catastrophic detonation at a BASF plant in Oppau, Germany, in 1921, which killed more than five hundred people and carved a crater ninety meters wide into the earth, happened because workers were trying to loosen a solidified fertilizer stockpile using dynamite, on the assumption that the material could not possibly explode. It could. It did. The town lost eighty percent of its housing stock in a single morning.

None of these stories are included for shock value, although they are, without embellishment, shocking. They are included because each one teaches something specific and transferable about how exothermic processes behave, and because understanding the mechanism behind a disaster is the only reliable way to prevent the next one. This is not a book about fear. Fear is a poor long term strategy for safety, because fear fades with familiarity and because it tends to attach itself to the wrong targets, the dramatic and cinematic dangers rather than the quiet, cumulative ones. What this book offers instead is something closer to literacy. By the time you finish it, you will understand why a pile of damp hay can reach two hundred degrees Fahrenheit without any external ignition source at all, why pouring water on a grease fire is one of the worst decisions a person can make in a kitchen, why lithium ion batteries fail in a specific and predictable sequence rather than randomly, and why the same chemical property that makes thermite useful for welding railroad tracks together also makes it functionally useless to fight with a garden hose.

The organization of the book follows a deliberate logic. The opening chapter builds the conceptual foundation, explaining in plain language what exothermic actually means, how the concept was developed by chemists working more than two centuries ago, and why the popular image of an exothermic reaction as something fast and dramatic captures only a narrow slice of what the term actually covers. From there, the book moves chapter by chapter through specific domains where exothermic chemistry creates real and often underappreciated hazards: agricultural fires caused by fermenting hay and grain, the counterintuitive violence of reactive metals meeting water, the concentrated heat of thermite and its long and complicated relationship with railroads, warfare, and popular imagination, the mechanics of dust explosions in grain elevators, the specific and devastating chemistry of ammonium nitrate, the anatomy of industrial reactor failures, the thermal runaway that occurs inside a damaged lithium ion battery, the physics of a kitchen grease fire, the aging and increasing instability of organic peroxides sitting forgotten in storage, and the modern laboratory science of predicting and preventing all of the above before it happens. A closing chapter draws these threads together into something practical: a way of thinking about exothermic risk that neither dismisses it nor catastrophizes it, but treats it the way an experienced electrician treats a live wire, with respect, procedure, and an accurate mental model of what is actually going on beneath the surface.

Throughout, the guiding conviction is that good information, clearly explained, is itself a form of protection. Farmers who understand the moisture threshold at which stored hay becomes a fire risk can check for it. Facility managers who understand the specific chemistry of ammonium sulfate contamination can avoid replicating the exact conditions that destroyed Oppau. Home cooks who understand why water and burning oil are a catastrophic combination will reach for a lid or a fire blanket instead, a choice that has saved kitchens and lives more times than statistics can easily capture. None of this requires a chemistry degree. It requires only the willingness to look past the deceptive calm of something at room temperature and ask the more useful question: what is this thing actually doing, chemically, right now, while it sits there looking like nothing at all.

It is worth being explicit, at the outset, about what kind of book this is and is not, since the subject matter invites two very different, equally unhelpful approaches that this book has deliberately tried to avoid. It is not a survivalist's manual, promising an exhaustive catalogue of every possible chemical danger a reader might one day encounter, presented in a register calculated to maximize anxiety about an unpredictable and hostile material world. Readers looking for that kind of encyclopedic dread will not find it here, and the chapters that follow are, if anything, considerably more interested in explaining why most of the substances discussed in this book are perfectly safe under ordinary conditions than in cataloguing every conceivable way they might fail. Nor is this a textbook, in the sense of proceeding through the material with the systematic, exam oriented thoroughness of a university chemistry course, working methodically through every relevant equation and every documented exception before allowing itself to tell a story. This is, instead, a book organized around narrative and consequence, built from the conviction that a small number of well chosen, well researched, real world stories, told with genuine attention to both the human and chemical details involved, teach the underlying science more durably, and considerably more enjoyably, than a purely systematic treatment ever could.

This narrative approach carries a specific implication worth stating plainly before the story properly begins. Every incident recounted in the chapters that follow is true, drawn from published investigative reports, peer reviewed scientific literature, and documented historical record, and the book has taken considerable care, throughout, to represent the specific facts of each case, casualty figures, chemical mechanisms, regulatory timelines, as accurately as the available public record allows. But truth, in the case of a technical and historical subject this wide ranging, is not the same thing as completeness, and readers with a specialized professional interest in any single incident discussed here, a particular disaster, a particular chemical hazard, a particular regulatory history, will find considerably more depth available in the specialized literature this book draws upon and, in its closing pages, points toward directly. What this book offers instead is breadth organized around a single, genuinely unifying scientific idea, the behavior of exothermic reactions across an unusually wide range of settings, industrial, agricultural, domestic, and technological, presented with enough narrative and human detail that the underlying chemistry, rather than sitting inertly on the page the way it too often does in more conventional treatments of the subject, actually sticks.

There is a further point worth making about who tends to encounter these hazards in practice, since it bears directly on why this book was written the way it was. The people most exposed to the dangers examined in these pages are rarely professional chemists, and rarely work in settings with ready access to a dedicated chemical safety officer or an industrial hygienist on staff. They are farmers checking on stored feed, home cooks managing a stovetop, delivery workers charging the battery that powers their livelihood, graduate students handling an unfamiliar compound for the first time, and shift workers arriving partway through a process someone else began earlier that day. Each of these people is fully capable of understanding the relevant chemistry, given a clear enough explanation, but very few of them have any obvious occasion to seek out a chemical engineering textbook or a government investigation report on their own initiative, nor should they need to in order to stay reasonably safe in the ordinary course of their work and daily lives. This book was written on the conviction that the distance between these readers and the specific, well established knowledge that could protect them is a solvable problem rather than a permanent one, and that closing even part of that distance, one reader and one clearly explained mechanism at a time, justifies the considerable effort a project of this scope requires.

The ice, as it turns out, was never really cold. It only looked that way from the outside.

Chapter 1

The Grammar of Heat

Before any of the disasters in this book can make sense, one word needs to be properly unpacked, because almost everyone has heard it and almost nobody uses it precisely. The word is exothermic, and its dictionary definition, a reaction that releases energy in the form of heat, is correct but nearly useless on its own. It tells you what happens without telling you why it happens, how fast it might happen, or what determines the difference between a reaction that warms a room by two degrees and one that levels a building. This chapter is an attempt to give the word its full meaning back.

Start with the basic bookkeeping of a chemical reaction. Every substance holds a certain amount of energy locked inside its chemical bonds, the connections between atoms that give a molecule its shape and identity. Breaking a bond costs energy. Forming a new bond releases energy. A chemical reaction is, at its core, nothing more than an accounting exercise in which old bonds are broken and new bonds are formed, and the question of whether the overall process is exothermic or endothermic comes down to a simple comparison: does forming the new bonds release more energy than breaking the old ones cost? If the answer is yes, the surplus energy has to go somewhere, and in the overwhelming majority of everyday chemical reactions, it goes out into the surrounding environment as heat. That is exothermic. If the answer is no, if forming the new bonds releases less energy than breaking the old ones required, the reaction has to pull energy in from its surroundings to make up the difference, and the immediate area gets colder rather than warmer. That is endothermic, and it is the less common and generally less dangerous of the two categories, since most endothermic reactions simply stall out once the available heat runs low.

Chemists formalize this energy accounting using a quantity called enthalpy, usually written with the symbol H, and the change in enthalpy during a reaction, written as delta H, is the number that tells you whether a reaction gives off heat or absorbs it. A negative delta H means the products of the reaction contain less stored chemical energy than the starting materials did, and the missing energy has been released, typically as heat. A positive delta H means the reverse. The size of that number, not just its sign, matters enormously for anything discussed in this book. A reaction with a small negative delta H, releasing perhaps a few kilojoules per mole, will warm its surroundings gently and probably safely. A reaction with a large negative delta H, releasing hundreds of kilojoules per mole, is capable of raising temperatures fast enough and high enough to ignite nearby materials, melt metal, or in the most extreme cases, produce genuine explosive force. The thermite reaction discussed in a later chapter, in which aluminum metal reacts with iron oxide, releases roughly enough energy per gram to melt the iron it produces on the spot, reaching estimated temperatures above two thousand degrees Celsius. That is not a difference of degree from a rusting nail. It is a difference of scale so large that describing both as simply exothermic risks obscuring more than it reveals.

This is where a second concept becomes essential, one that is arguably more important to the subject of this book than enthalpy itself: activation energy.

Knowing that a reaction will release energy overall tells you nothing about whether it will actually happen, or how easily it can be started. Activation energy is the energy barrier that has to be overcome before a reaction can begin, regardless of how much energy it eventually releases. Think of a boulder resting in a shallow depression at the top of a hill. The boulder wants, in an energetic sense, to roll down the hill, releasing a great deal of gravitational energy as it does. But it cannot do so until it is nudged out of the depression that is currently holding it in place. That nudge is the activation energy. A reaction can be wildly exothermic overall and still sit at room temperature indefinitely, completely stable, simply because nothing has provided the initial nudge required to get it started. This is precisely the situation with a great many of the substances that appear later in this book. Ammonium nitrate fertilizer is thermodynamically capable of releasing enormous amounts of energy, but under ordinary conditions its activation energy is high enough that it sits in warehouses and on farms across the world, year after year, without incident. It is only when something lowers that barrier, contamination with combustible material, an intense and sustained fire, extreme confinement, that the boulder gets its nudge, and by then it is generally too late to stop what follows.

The interplay between these two quantities, the overall energy released and the barrier required to start releasing it, explains almost every counterintuitive fact in this book. It explains why a bottle of diethyl ether can sit safely on a shelf for years and then become dangerously shock sensitive as it ages, because the slow formation of peroxide compounds within the ether gradually lowers the activation energy required to trigger a violent decomposition, even as the bottle looks and smells exactly the same as it did the day it was purchased. It explains why lithium ion batteries can be handled safely for years of ordinary use and then fail catastrophically after a single sharp impact, because mechanical damage creates an internal short circuit that effectively hands the reaction its activation energy directly, all at once, in a location with nowhere for the resulting heat to escape. And it explains why some of the most dangerous exothermic processes in this book are not fast reactions at all, but slow ones that generate just enough heat to gradually erode their own activation energy barrier, a phenomenon with its own name and its own chapter later in this book: thermal runaway.

Thermal runaway deserves an early and careful introduction here because it is, in a sense, the organizing idea behind more than half of the case studies that follow. Most people who have taken a chemistry class understand reaction rate as something that increases with temperature, and generally understand this in a vague, linear way, warmer means faster. The real relationship, described by the Swedish chemist Svante Arrhenius in the 1880s in an equation that now bears his name, is not linear at all. It is exponential. A modest increase in temperature, ten degrees Celsius in many common reactions, can roughly double the rate at which a reaction proceeds. Double the rate means, all else being equal, double the rate of heat production. If that heat is not removed as quickly as it is generated, the temperature climbs further, which doubles the reaction rate again, which increases heat production again, in a feedback loop that accelerates every cycle through it. This is thermal runaway: not a single dramatic event, but a self reinforcing spiral in which heat and reaction rate escalate each other until something fails, whether that something is a sealed industrial reactor, as at T2 Laboratories in 2007, or the internal structure of a lithium ion battery cell, or a haystack that has been quietly self heating in a barn for three weeks.

It is worth pausing on how old this science actually is, because one of the persistent myths surrounding chemical hazards is that they are somehow a modern problem, a byproduct of industrial complexity that previous generations did not have to reckon with. The opposite is closer to the truth. The formal study of reaction heat dates back to the French chemist Antoine Lavoisier, working in the 1780s, who along with the mathematician Pierre Simon Laplace built one of the first calorimeters, a device that measured heat by tracking how much ice it could melt during a chemical or biological process. Lavoisier's work established that combustion and respiration were fundamentally the same kind of process, both consuming oxygen and releasing heat, a genuinely radical idea at the time and one that laid the groundwork for everything that followed in thermochemistry. The Swiss born chemist Germain Hess, working in Russia in the 1840s, demonstrated that the total heat released or absorbed in a chemical reaction is independent of the specific pathway taken to get from starting materials to final products, a principle now known as Hess's Law, which remains one of the load bearing rules of modern thermochemistry and which allows chemists to calculate the heat of reactions that would be far too dangerous to actually run in a laboratory and measure directly. Later in the nineteenth century, the French chemist Marcellin Berthelot conducted extensive calorimetric studies of explosive reactions specifically, in part because understanding the energy release of explosives was of obvious military and industrial interest, and his measurements remain a foundational reference in the field of energetic materials to this day.

None of this is ancient history in the sense of being irrelevant. It is ancient history in the sense of being extremely well established. The chemistry that caused the West Fertilizer Company disaster in 2013 was not some newly discovered hazard that caught scientists by surprise. Ammonium nitrate's capacity for violent decomposition under the right conditions had been documented since at least the Oppau disaster of 1921, and arguably long before that. What fails, again and again, in the case studies that make up this book, is not the science. The science is solid, has been solid for well over a century, and is not particularly difficult to explain to a general audience, which is part of the reason this book exists. What fails is the translation of that science into the judgment of the people standing next to the reactor, the grain bin, the fertilizer stockpile, or the frying pan, in the moment when a decision has to be made. Understanding delta H and activation energy will not by itself keep anyone safe. But it does something that turns out to matter a great deal in practice: it changes what a person notices. A person who understands thermal runaway does not see a slightly warm compost pile and think nothing of it. They see a system that is currently accelerating, and they understand why that acceleration, left unaddressed, tends to end badly.

It is worth spending a little more time with Arrhenius himself, since the equation that bears his name is arguably the single most load bearing piece of mathematics in this entire book, quietly underwriting every chapter that follows even in chapters that never mention his name again. Arrhenius was, by most biographical accounts, a genuine child prodigy, reportedly teaching himself to read at the age of three by watching his father, a land surveyor for Uppsala University, work through columns of figures in an account book. He entered university at seventeen and, notably, nearly failed to have his doctoral dissertation accepted at all, receiving a grade so lukewarm from his examining committee that it barely qualified as a pass, largely because his theories about ions and electrical conductivity in solution struck the chemistry establishment of his day as implausible. That same body of work later earned him the Nobel Prize in Chemistry in 1903, the first ever awarded to a Swedish scientist, a reminder that the distance between an unpopular new idea and an eventually foundational one is not always obvious to the people evaluating it at the time.

The equation that concerns this book specifically, describing how reaction rate depends on temperature, was not entirely original to Arrhenius. The Dutch chemist Jacobus Henricus van 't Hoff had already proposed a mathematical relationship along similar lines in 1884, working from observations about how equilibrium constants shifted with temperature. What Arrhenius contributed, in a paper published in 1889, was the physical interpretation that gave the mathematics its lasting significance: the insight that the equation's structure could be explained by imagining that only a fraction of molecules in any given sample possess enough energy, at any given moment, to actually react when they collide, and that this fraction grows rapidly, non-linearly, as temperature rises. Arrhenius gave this energy threshold a name, activation energy, and the equation that bears his name expresses, with remarkable mathematical economy, exactly how sensitively that fraction of sufficiently energetic molecules responds to even modest changes in temperature. It is this equation, more than any other single piece of chemistry, that explains why a ten degree Celsius increase can roughly double a reaction's rate, why a hay stack that sits safely at ambient temperature for weeks can suddenly begin accelerating toward ignition once internal heating crosses a threshold, and why industrial engineers treat every additional few degrees of uncontrolled temperature rise inside a reactor as a meaningfully more urgent problem than the last, rather than as a simple linear progression toward danger.

There is a second, quieter reason to dwell on the history of thermochemistry before moving into this book's case studies, and it concerns the difference between a well established scientific principle and a widely internalized one. Every quantity introduced in this chapter, enthalpy, activation energy, the exponential temperature dependence of reaction rate, has been part of the standard chemistry curriculum for the better part of a century. None of it is cutting edge, contested, or awaiting further confirmation. And yet, as later chapters of this book will demonstrate in considerable and sometimes tragic detail, a settled scientific principle and a principle that is actually, reliably applied by the people whose safety depends on it are two very different things. A concept can sit, fully verified and thoroughly taught, in textbooks for a hundred years and still fail to travel the short remaining distance into the specific judgment of a specific person standing next to a specific reactor on a specific afternoon. That gap, between the science being known and the science being used, is in many ways the true subject of this book, and the grammar introduced in this chapter, delta H, activation energy, and the Arrhenius relationship binding temperature to rate, is simply the vocabulary needed to see that gap clearly wherever it appears in the chapters ahead.

It is also worth being precise, this early in the book, about what this chapter's vocabulary does and does not predict, since a common and understandable misconception is to assume that a reaction's danger can be read directly off its enthalpy value alone, the bigger the negative delta H, the more hazardous the substance. This is not quite right, and the distinction matters enormously for how the rest of this book should be read. A reaction with an enormous negative delta H but a very high activation energy, meaning it releases tremendous energy once started but is extremely difficult to start in the first place, can be entirely safe to store, handle, and transport under normal conditions, precisely the situation with thermite, examined in a later chapter, which sits inert and stable through years of storage and transport specifically because its activation energy is high enough to exclude accidental ignition from ordinary sparks or friction. Conversely, a reaction with a comparatively modest negative delta H but a low or steadily eroding activation energy can become dangerous specifically because it is so easy to trigger, or because something in its storage conditions is actively working to lower that barrier over time, precisely the situation with aging organic peroxides, examined in a later chapter, where years of slow autoxidation gradually transform a stable solvent into a compound capable of detonating from nothing more than the mechanical shock of a twisted bottle cap. True hazard, in other words, is always a function of both quantities considered together, the size of the energy release and the ease with which that release can be triggered, and one of this book's central organizing tasks, chapter by chapter, is tracing exactly how each specific substance and setting combines these two variables into its own particular, sometimes counterintuitive risk profile.

The remaining chapters of this book are, in a sense, an extended demonstration of these two ideas, delta H and activation energy, playing out across an enormous range of real and different settings, from a Texas fertilizer warehouse to a New York City apartment building to a nineteenth century German railroad yard to a Soviet nuclear reactor operating under conditions its own designers had not fully anticipated. The chemistry, and in one case the underlying nuclear physics operating according to a directly analogous logic, is consistent throughout, even when the settings and the substances vary wildly. What changes from chapter to chapter is the specific mechanism by which a reaction that has been sitting quietly, apparently inert, finally gets the nudge it needed. Sometimes that nudge is moisture. Sometimes it is friction, or a spark, or simple time, as aging chemicals slowly lower their own activation energy through processes the original manufacturer never intended and rarely warned about. Understanding the grammar of heat, the vocabulary of enthalpy, activation energy, and runaway feedback, is what allows all of these otherwise disconnected stories to be read as chapters of a single, coherent subject rather than a collection of unrelated tragedies. They are not unrelated. They are variations on one theme, and that theme is the subject of everything that follows.

Chapter 2

The Slow Burn: Spontaneous Combustion in Hay, Coal, and Compost

A farmer who loses a barn to fire is statistically far more likely to have lost it to the hay inside than to any wiring fault, lightning strike, or careless cigarette. This fact surprises most people outside of agriculture, and it surprised a good many people inside agriculture too, for a long stretch of the twentieth century, because the mechanism behind it does not look anything like what people picture when they picture a fire starting. There is no spark. There is no obvious source of ignition at all. There is only hay, stacked and left alone, quietly heating itself from the inside until it reaches a temperature at which it ignites on its own. This chapter is about that process, spontaneous combustion, and about why it remains, even now, one of the most underappreciated fire hazards in agriculture, industry, and the ordinary household garage.

The mechanism begins, somewhat counterintuitively, with biology rather than chemistry. Hay that is baled or stored with too much residual moisture, generally above twenty five percent moisture content, becomes an active habitat for bacteria and fungi almost immediately after storage. These organisms feed on the plant material and, like all living things engaged in metabolism, they generate heat as a byproduct. In a small, well ventilated pile, this heat simply escapes into the surrounding air and nothing further happens. But hay is rarely stored in small, well ventilated piles. It is stored in large, dense stacks and bales specifically because density is efficient for storage and transport, and that same density is precisely what traps the heat the microorganisms are generating. As the interior of the stack warms, somewhere in the range of one hundred thirty to one hundred forty degrees Fahrenheit, the microbial activity that started the process actually begins to die off, because most of the bacteria and fungi involved cannot survive at those temperatures. If the story stopped there, spontaneous combustion would not exist as a hazard, because the heat source would simply shut itself off.

The story does not stop there. Above roughly one hundred thirty degrees Fahrenheit, a second and entirely different process takes over: chemical oxidation. The organic compounds within the hay, sugars, plant fibers, and residual moisture, begin reacting directly with atmospheric oxygen in a slow combustion process that requires no living organism to sustain it. This chemical oxidation is itself exothermic, releasing its own heat, and because it does not depend on bacteria that die off at high temperature, it can continue accelerating well past the point where the biological process quit. This is the handoff that makes hay fires so persistently mysterious to people encountering the phenomenon for the first time: a process that starts as biology becomes a process driven by pure chemistry, and the chemistry, following exactly the exponential relationship between temperature and reaction rate described in the previous chapter, accelerates itself. As the interior of a hay stack climbs past one hundred seventy five degrees Fahrenheit, and eventually toward two hundred forty degrees, the material can begin to char, and charred plant material has a dramatically lower ignition temperature than the original hay. Somewhere in the range of two hundred forty to two hundred eighty degrees Fahrenheit, the smoldering core of a hay stack that has been silently building heat for anywhere from one to eight weeks can finally break into open flame, at which point the fire, once it reaches oxygen from an air pocket or the surface of the stack, can spread with startling speed.

Agricultural extension services in the United States, Canada, and Europe have studied and documented this process for well over a century, and the practical guidance that has emerged from that research is remarkably consistent and specific. Hay intended for long term storage should generally be baled at a moisture content below twenty percent, and material above forty percent moisture is considered to be entering a genuinely dangerous range for spontaneous heating. The timing matters as much as the moisture level. Field studies and farm loss records consistently show that the highest risk window for a spontaneous hay fire falls within the first two months of storage, and more specifically within the fourth to sixth week, which lines up precisely with the timeline required for the microbial heating phase to peak and hand off to chemical oxidation. Farmers who monitor hay temperature during this window, using a simple probe thermometer inserted into the stack, are following a practice that has changed remarkably little in a hundred years, because the underlying chemistry has not changed either. A reading above one hundred seventy five degrees Fahrenheit is generally treated as cause for serious concern, and readings that continue climbing rather than stabilizing are treated as a signal to remove the material from the structure entirely, since a burning haystack pulled outside a barn causes vastly less damage, and puts vastly fewer lives at risk, than the same fire breaking out inside one.

The economics of this hazard are worth understanding as well, because they explain why the risk has never been fully engineered away despite being well understood for so long. Storing hay densely is not a mistake or an oversight, it is the only economically sensible way to store it, since loose, well ventilated hay storage would require dramatically more building volume per ton of feed, an unaffordable proposition for most working farms. A single silo holding several hundred tons of corn silage represents a genuinely large capital investment in feed, one that a farm operation often cannot simply replace if it is lost, and the incentive to store that feed as densely and cheaply as possible is therefore very strong, working directly against the ventilation and monitoring practices that reduce fire risk. This is a recurring pattern throughout this book: the conditions that create exothermic hazard are frequently the same conditions that make economic sense, which is precisely why the hazard persists generation after generation despite being, in a narrow technical sense, entirely preventable.

Hay is the most extensively documented case of agricultural spontaneous combustion, but it is far from the only one, and understanding the broader category helps clarify why this chapter belongs early in a book that otherwise ranges into industrial chemistry, batteries, and kitchen fires. Stockpiled coal is subject to a closely related process, in which coal particles oxidize slowly in contact with air, generating heat that becomes trapped within a large pile in much the same way heat becomes trapped within a dense hay stack. Coal stockpile fires have historically been a significant concern at power generation facilities and mining operations, particularly for coal with higher sulfur or moisture content, which oxidizes more readily. The practical countermeasures mining and utility operators use, compacting piles to exclude air rather than allowing it to circulate, limiting pile height, monitoring internal temperature, and rotating older stock to prevent long term accumulation, are conceptually the same countermeasures used against hay fires, applied to an entirely different material because the underlying chemistry, slow oxidation generating heat faster than it can escape, is the same.

No case illustrates the potential consequences of an unaddressed coal self heating event more vividly, or more durably, than the fire that has burned beneath the small town of Centralia, in the anthracite coal region of eastern Pennsylvania, continuously since 1962. The fire's origin was, on its surface, entirely mundane and entirely unrelated to any of the industrial hay or coal pile storage scenarios described above. In the spring of 1962, the Centralia Borough Council authorized a routine cleanup of an abandoned strip mining pit at the edge of town that had been used for some time as an unregulated landfill, and the standard method for reducing the volume of trash accumulated there was simply to set it alight. That fire, however, found its way into an exposed seam of anthracite coal concealed behind the landfill material, and coal, once ignited underground within a seam that offers a sustained supply of oxidation surface and only limited pathways for the resulting heat to escape, behaves according to precisely the same self accelerating thermal dynamic this chapter has already described in stockpiles and stored hay. Research into the underlying combustion chemistry indicates that for roughly every eighteen degree Fahrenheit rise in coal temperature, the reaction rate driving further oxidation can double, an exponential relationship that will be immediately recognizable to readers of this book's opening chapter as the same Arrhenius dynamic governing every runaway exothermic process examined throughout these pages, here operating across a scale of decades rather than the weeks typical of a hay barn or the seconds typical of a battery cell.

The Centralia fire, once established within the coal seam, proved essentially unstoppable using the resources and techniques available at the time, and it has since spread through passageways beneath an area estimated at several hundred acres, by some later surveys touching land beneath as much as four hundred acres of the town and its surroundings. The fire was not fully appreciated as an existential threat to the town until 1979, when a local gas station owner discovered, upon inserting a dipstick to check fuel levels in an underground storage tank, that the gasoline registered a temperature of 172 degrees Fahrenheit, an alarming reading that prompted more serious investigation into just how far the underground fire had spread and how close it had come to the town's inhabited areas. The situation reached a definitive crisis point in 1981, when a twelve year old resident of the town, playing in his grandmother's backyard, fell into a sinkhole that opened suddenly beneath him, a collapse caused by the fire's slow undermining of the ground above the burning coal seam, and one that released a visible plume of steam and carbon monoxide gas from the opening, a near fatal reminder that the fire's danger extended well beyond property damage to encompass genuine and immediate risk to human life from both structural collapse and toxic gas exposure. In the years that followed, the federal government allocated relocation funding, and the great majority of

Centralia's roughly one thousand residents accepted buyouts and left permanently, reducing what had been a functioning small town to what current accounts describe as a near ghost town, home to only a small handful of remaining residents who negotiated the right to stay for the remainder of their lives, surrounded by cracked and abandoned streets, demolished foundations, and a fire that, more than sixty years after that spring cleanup burn in 1962, continues to smolder beneath the ground, by some estimates possessing enough remaining coal fuel to continue burning for another century or more.

Centralia is, by a wide margin, the most famous of an entire, sobering category of ongoing underground coal fires, a category that includes more than two hundred currently active fires within the United States alone according to surveys conducted in the early 2010s, and considerably larger numbers across major coal producing regions internationally, including well documented, long burning fires across coalfields in China, India, and South Africa. The specific engineering and regulatory response to this broader hazard, sealing exposed coal seams from atmospheric oxygen wherever practical, monitoring stockpile and mine temperatures using the same basic thermal surveillance principles described earlier in this chapter regarding hay, and in modern coal handling operations, applying specialized sealant coatings to stockpiles specifically formulated to reduce the surface oxidation that initiates self heating, mirrors closely the countermeasures already described for hay and grain, reflecting once again the underlying unity of this book's subject matter: different fuels, different settings, and vastly different timescales, all governed by the same handful of thermodynamic principles introduced in this book's opening chapter, and all defeated, when they are defeated at all, by the same basic combination of moisture control, oxygen exclusion, and vigilant, sustained temperature monitoring.

Compost presents a gentler but structurally identical version of the same phenomenon, and it is worth including here because so many people manage a compost pile in their own backyard without ever considering that they are, in a small way, running the same kind of process that burns down barns. Composting is intentionally designed to generate heat, since the heat produced by microbial breakdown of organic material is what accelerates decomposition and, at sufficiently high sustained temperatures, kills weed seeds and pathogens, which is why well managed compost piles are actively encouraged to reach one hundred thirty to one hundred sixty degrees Fahrenheit during their active phase. Most home composting never proceeds past this biological stage into genuine fire risk, largely because home compost piles tend to be smaller, less densely packed, and more frequently turned than a hay stack or coal pile, all of which allow trapped heat to dissipate. But larger scale composting operations, particularly those handling high volumes of nitrogen rich material like grass clippings or manure in large windrows, have experienced genuine spontaneous combustion incidents for the same underlying reason a barn does: enough mass, enough insulation, and not enough turnover of air to let the heat escape before chemical oxidation takes over from biology.

The common thread running through hay, coal, and compost is worth stating plainly, because it is the thread that connects this chapter to nearly everything that follows in the rest of the book. None of these materials require an external ignition source in the conventional sense. There is no match, no spark, no faulty wire. The ignition source is internal, generated by the material's own chemistry, and it builds gradually enough that by the time it becomes visible, smoke rising from the top of a hay stack, an unusual smell coming from a coal pile, the process is often already well advanced and difficult to stop by any means short of physically breaking apart the pile to expose its hot core to open air, a maneuver that itself risks providing the oxygen needed to convert smoldering into open flame. This is precisely the character of exothermic danger this book is most concerned with: not the sudden, obvious event, but the slow, cumulative one that looks, right up until it does not, exactly like nothing at all.

There is a final point worth making about why this particular hazard remains persistently underestimated, and it has less to do with chemistry than with human psychology, a subject this book returns to in its closing chapter. A fire that starts from an obvious external cause, a lightning strike, a dropped cigarette, an electrical short, fits neatly into a mental category most people already possess: fires are caused by something. A fire that emerges from the material's own quiet internal chemistry does not fit that category comfortably, and the mind tends to resist explanations that do not match its existing framework. Farmers who have never personally experienced a hay fire sometimes treat the moisture and temperature guidance from extension services as an abundance of caution rather than a hard chemical limit, in part because the danger genuinely does not look like danger while it is developing. It looks like a stack of hay sitting quietly in a barn, doing nothing at all, which is exactly what it will continue to look like, right up until the morning it does not.

The practical monitoring guidance that agricultural extension services and farm insurers have converged on over decades of accumulated experience is worth setting out in some detail, because it translates the abstract chemistry described above into a specific, actionable protocol that any farm operation can follow without any specialized equipment beyond a length of pipe and a thermometer. The process begins before the hay is ever baled, since moisture content at the moment of baling is the single strongest predictor of whether a stack will develop a dangerous internal heat source at all. Agricultural guidance generally recommends baling small rectangular bales only when moisture content tests at 20 percent or below, with a slightly stricter threshold of 18 percent for large rectangular or round bales, which pack more densely and therefore trap heat more effectively once stacked. Farmers who bale without access to an electric moisture tester, a relatively inexpensive handheld device that measures moisture content directly, have traditionally relied on judging conditions by feel and by weather, since hay cured during a stretch of windy, low humidity weather dries more reliably and more evenly than hay baled during humid conditions or under a heavy morning dew, and many experienced farm operators specifically avoid baling too early in the day for exactly this reason, since dew moisture picked up overnight can push an otherwise adequately cured field over the dangerous threshold.

Once hay is baled and stacked, the monitoring obligation shifts from moisture to temperature, and the practical technique farmers have used for generations, refined only modestly by modern instrumentation, is elegantly simple. A probe, either a purpose built agricultural thermometer with a sufficiently long stem or a home fabricated version made from several feet of narrow iron pipe with a pointed tip and a thermometer lowered inside on a string, is driven deep into the interior of the stack, since surface temperature readings reveal essentially nothing about the hidden, insulated core where dangerous heating actually develops. Extension guidance generally recommends checking newly stacked hay at least once, and ideally twice, daily for the first six weeks after baling, the window during which the biological to chemical handoff described earlier in this chapter is most likely to occur. The specific readings farmers are trained to watch for form a graduated response scale that has remained remarkably consistent across different extension services and insurers: at 130 degrees Fahrenheit, the hay is beginning active heating and should be monitored closely; at 160 degrees, checks should increase to every four hours and physical inspection of the stack should begin; at 175 degrees, the hay should be moved away from any structure if this can be done safely, and a fire department should generally be notified; and above that threshold, standard guidance uniformly cautions against moving or disturbing the hay at all, since introducing fresh air to a smoldering, oxygen starved core is exactly the kind of intervention that can convert a contained, slow burning problem into an open, fast spreading fire, a point of practical wisdom that mirrors, in a strikingly direct way, the broader lesson about oxygen and confinement that runs throughout later chapters of this book concerning dust explosions and reactor safety.

A 2024 incident at an animal sanctuary in Talsarn, in west Wales, illustrates how closely this protocol, when followed, can catch a developing hazard before it becomes a catastrophe, and also how quickly the situation can escalate even after monitoring has begun. Local fire and rescue services responded to a report that hay bales stored at the property felt warm to the touch, a report that in itself represents exactly the kind of early, informal warning sign this chapter has emphasized throughout. Responding officers tested the bales for both heat and moisture content and returned later the same day to continue monitoring, at which point twenty bales were removed from the barn as a precaution. On reinspection the following day, temperatures had climbed to a level investigators characterized as critical, and approximately one hundred bales ultimately had to be removed from the structure, with smoke visibly issuing from some of them by the time crews intervened, a sequence that moved from a mild warm to the touch report to active smoldering requiring substantial physical intervention within roughly twenty four hours, underscoring both how quickly a self heating hay stack can escalate once it crosses a certain threshold and how much protective value even a delayed but attentive response can still provide, since the sanctuary in this case lost stored feed but not, as far as the public incident record indicates, the structure housing it or any animals within it.

This kind of near miss, common enough that fire services in agricultural regions of the United Kingdom, the United States, and Canada routinely publicize both the hazard and, in some cases, offer free bale temperature testing services to local farmers, illustrates precisely the value proposition at the heart of this chapter's argument. The protocol described above requires no chemistry degree, no expensive instrumentation, and no fundamental change to how hay is farmed, baled, or stored. It requires only the discipline of checking, at predictable intervals, during a predictable and well documented window of vulnerability, using tools that have barely changed in a century. The chemistry of spontaneous combustion in hay has not, in any meaningful sense, gotten more dangerous over that century. What has occasionally lapsed, farm by farm and season by season, is the consistency with which that simple, well established checking routine actually gets performed, a pattern of occasional lapsed vigilance around a well understood hazard that will recur, in different guises and different industries, throughout the remaining chapters of this book.

Chapter 3

When Water Is the Enemy: Reactive Metals and the Sodium Myth

Water is the substance human beings use to fight fire more often than any other, so completely and so instinctively that the association feels less like a learned fact and more like a law of nature. This makes the subject of this chapter genuinely difficult for most readers to accept on a gut level, even after the chemistry has been explained clearly: there exists a category of materials for which water is not a solution but the trigger itself, the precise thing that turns a stable, inert lump of metal into a violent, self sustaining reaction. Understanding why requires setting aside the assumption that water is chemically passive, because for a specific and important group of elements, it is anything but.

The alkali metals, the leftmost column of the periodic table, include lithium, sodium, potassium, rubidium, and cesium, and they share a defining chemical trait: each has a single loosely held electron in its outermost shell, an electron the atom is, in a chemical sense, eager to give away. Water, meanwhile, is a molecule that is happy to accept an electron under the right conditions, and when an alkali metal contacts water, that transfer happens essentially immediately. The reaction produces a metal hydroxide, hydrogen gas, and a substantial release of heat, and it is this combination, heat and flammable hydrogen gas appearing together in the same small space, that makes the reaction so visually dramatic. Sodium dropped into a beaker of water in a chemistry classroom will hiss, dart across the surface, and often ignite the hydrogen it produces, sometimes with a small explosion. Potassium, one row further down the periodic table and therefore holding its outer electron even more loosely, reacts faster and more violently still, typically igniting immediately upon contact. This escalation continues down the column, and it is precisely the kind of memorable demonstration that has made alkali metal reactivity a staple of introductory chemistry education for well over a century, right alongside its considerable practical hazard whenever these metals are handled, stored, or disposed of outside a carefully controlled classroom setting.

For most of that century, the standard textbook explanation for why the reaction is explosive rather than merely fast rested on a straightforward chain of cause and effect: heat from the reaction ignites the hydrogen gas being produced, and the resulting combustion is what people perceive as the explosion. This explanation is not wrong, but a research team led by the chemist Pavel Jungwirth at the Czech Academy of Sciences discovered, through a combination of high speed camera footage and molecular simulation published in the journal Nature Chemistry, that it was seriously incomplete, and that the true mechanism is considerably stranger. The team noted a problem with the conventional story that had apparently gone unaddressed for decades: as soon as an alkali metal begins reacting with water, it produces a layer of steam and hydrogen gas around itself, and that layer should act as a buffer, physically separating the remaining unreacted metal from further contact with water. By that logic, the reaction ought to smother itself quickly, the same way a fire starved of oxygen dies down, rather than accelerating into an explosion. And yet, unmistakably, it does not.

Using cameras capable of capturing thirty thousand frames per second, far faster than the human eye or any earlier study could resolve, Jungwirth's team filmed droplets of a sodium potassium alloy at the exact instant they contacted water, and what they found was that within a fraction of a millisecond, spikes of metal erupt outward from the surface of the droplet, physically piercing through the vapor layer that should have been protecting the remaining metal, and exposing fresh metal surface to fresh water again and again in rapid succession. Their molecular simulations offered an explanation for why these spikes form at all: the transfer of electrons from the metal into the surrounding water happens so quickly that the metal surface itself becomes positively charged, and the resulting electrical repulsion between the positive charges built up across the droplet's surface, a phenomenon the researchers described using a term borrowed from nuclear physics, a Coulomb explosion, physically tears the droplet apart from within, faster than any steam buffer can form to stop it. In other words, the explosion is not simply hydrogen catching fire after the fact. It begins as an electrical event, driven by the sheer speed and completeness with which alkali metals give up their outer electron, and only afterward does the resulting fragmentation and heat go on to ignite the hydrogen that has also been produced. This finding, published in the twenty first century about a reaction that chemistry students have performed and watched since the nineteenth, is a useful reminder that even the most classic, most thoroughly demonstrated chemical hazards can still contain mechanisms that were not fully understood until surprisingly recently.

The practical hazard scales predictably with the size of the metal involved and the setting in which the reaction occurs. A gram sized piece of sodium in a lecture hall demonstration, handled by a trained instructor with appropriate shielding, produces a startling but generally containable reaction. Industrial quantities of sodium, which is used at scale in applications ranging from certain nuclear reactor coolant systems, where liquid sodium's excellent heat transfer properties are valuable specifically because sodium remains liquid across a very wide temperature range, to chemical manufacturing and metal extraction processes, represent an entirely different order of hazard. Documented engineering literature on liquid sodium systems notes that the reaction between sodium vapor and water vapor becomes increasingly exothermic as sodium approaches and exceeds its boiling point of eight hundred eighty three degrees Celsius, and that explosions specifically become associated with sodium in a dispersed, high surface area state at that temperature, precisely the condition the Jungwirth research helps explain at the molecular level. Facilities that handle bulk quantities of reactive alkali metals accordingly maintain extensive protocols to prevent any contact with water or even atmospheric humidity, store the metal under inert oil or argon gas rather than in open air, and use specialized fire suppression methods, since standard water based firefighting equipment is, for this category of material, not merely ineffective but actively counterproductive, feeding the exact reaction it would be deployed to stop.

This last point, that water can be the wrong response to a fire rather than the default correct one, connects the alkali metal hazard to a considerably larger and more consequential story, and one that illustrates how the same basic principle, an exothermic reaction triggered or intensified by water contact, can operate at an industrial scale with catastrophic human cost even when the chemistry involved is entirely different from a sodium and water demonstration. On the night of December second into December third, 1984, at a pesticide manufacturing plant operated by Union Carbide India Limited in the city of Bhopal, a storage tank designated E610 held roughly forty tons of methyl isocyanate, a chemical intermediate used in the production of a common agricultural pesticide. Methyl isocyanate is, under normal conditions, a stable liquid, but it reacts vigorously and exothermically with water, and investigators established that on that night, water entered the tank, whether through a maintenance failure in the connected pipe system or, as an internal Union Carbide investigation later concluded, through deliberate introduction by a disgruntled employee, a conclusion that has remained disputed and contested in the decades since. What is not disputed is the chemistry that followed. The water triggered a runaway exothermic reaction inside the tank, and as the reaction progressed, temperature and pressure inside the sealed vessel climbed, slowly at first and then at an accelerating rate that mirrors precisely the thermal runaway dynamic described in the opening chapter of this book. The tank's refrigeration system, which was designed to keep the stored methyl isocyanate cold specifically to slow any reaction with contaminants, was not operating at the time. As pressure continued to build, the tank's safety relief valve eventually opened, and roughly forty tons of toxic gas vented directly into the densely populated neighborhoods surrounding the plant. Estimates of the resulting death toll vary depending on the timeframe and methodology used, with official Indian government figures documenting more than thirty seven hundred deaths in the immediate aftermath and independent estimates of total deaths from acute and longer term effects ranging considerably higher, alongside several hundred thousand people who suffered lasting respiratory, ocular, or other injuries.

Bhopal is, by any measure, the worst industrial disaster in recorded history, and it is included in this book not for its scale, which is well documented elsewhere and deserves treatment far more extensive than a single chapter can offer, but specifically for what it demonstrates about the subject of this book: an exothermic reaction, triggered by water contact with a reactive chemical, operating inside a sealed and inadequately monitored vessel, escalating from a containable problem into an uncontainable one within the span of about an hour. The same physics that makes a sodium demonstration dramatic in a classroom, an exothermic reaction with water producing heat and pressure faster than the surrounding system can safely dissipate it, scaled up by orders of magnitude in tank volume, chemical toxicity, and consequence, is what happened in Bhopal. The tank's designers understood that methyl isocyanate reacted with water. What failed, catastrophically, was every layer of the system meant to prevent that contact from occurring and to contain its consequences if it did.

There is a broader lesson embedded in placing these two cases side by side, the classroom sodium demonstration and the industrial catastrophe in Bhopal, and it is a lesson this book returns to repeatedly: the underlying chemical principle, in this instance the vigorous and often violent reactivity between water and certain classes of reactive chemicals, does not change based on scale or setting. What changes is consequence, and consequence is determined almost entirely by containment, monitoring, and the quality of the barriers placed between the reaction and the people nearby. A sodium demonstration performed in a fume hood behind safety glass, with a trained instructor and a small quantity of metal, is genuinely educational and genuinely low risk, not because the chemistry is different from the chemistry inside a bulk sodium storage tank, but because every element of the setting has been deliberately engineered to contain a small, predictable version of a reaction that would be extraordinarily dangerous at any larger scale or with less careful handling. Bhopal represents the opposite condition: a large quantity of a hazardous, water reactive chemical, stored in a system whose safety mechanisms had been allowed to degrade, situated immediately adjacent to a dense residential population with essentially no buffer zone. The chemistry itself offered no warning beyond what it had always offered. It was, as it had always been, exothermic, water reactive, and entirely indifferent to how carefully or carelessly it was being managed on any particular night.

A third case, sitting in scale somewhere between a classroom demonstration and a catastrophe like Bhopal, illustrates just how much engineering discipline is required to keep bulk quantities of a reactive alkali metal genuinely safe over years of continuous industrial use, and what happens on the occasions that discipline momentarily fails. Liquid sodium's physical properties, an exceptionally wide temperature range across which it remains liquid and an unusually high thermal conductivity, roughly one hundred twenty times that of water according to engineering literature on the subject, make it an outstanding coolant for a specific and demanding application: fast breeder nuclear reactors, a reactor design that operates without the extensive neutron moderation used in conventional water cooled reactors and that, for reasons of reactor physics beyond the scope of this book, produces or breeds additional nuclear fuel as it operates. Since the 1950s, nations including the United States, France, the United Kingdom, the Soviet Union, and Japan have built and operated sodium cooled fast reactors, accumulating, by some engineering estimates, several hundred combined reactor years of operating experience with liquid sodium systems, and that accumulated experience includes a recurring, well documented category of incident: the sodium leak.

Japan's Monju prototype fast breeder reactor, a two hundred eighty megawatt facility near Tsuruga that began operation in 1994, provides one of the most thoroughly documented examples. During performance testing in December 1995, a thermometer sensor housing installed in a secondary heat transport pipe failed due to metal fatigue caused by vibration from the flowing sodium itself, a comparatively small and, on paper, unremarkable mechanical component whose failure nonetheless allowed an estimated six hundred forty kilograms of high temperature liquid sodium to escape through the resulting breach. Sodium exposed to air ignites essentially on contact, precisely the reactivity mechanism examined earlier in this chapter, and the leaking sodium at Monju caught fire almost immediately upon reaching the surrounding air, ultimately consuming roughly three tonnes of sodium in the resulting blaze before the situation was brought under control. Investigators later concluded that the reactor core itself had remained safely cooled throughout the incident and that no radiological hazard to the surrounding environment or public had resulted, a genuinely important distinction between a sodium fire, a serious industrial hazard in its own right but chemically contained to the sodium itself, and a nuclear criticality or containment failure, the far more severe category of incident examined in a separate chapter of this book. Even so, the Monju sodium fire proved enormously consequential in a different sense entirely: the plant's operator was later found to have mishandled the incident's immediate aftermath, including editing video footage of the accident before releasing it to investigators and delaying formal notification of surrounding communities by roughly an hour, a breakdown in transparency that provoked such sustained public and regulatory backlash that the reactor remained shut down for very close to fifteen years before eventually resuming limited operation.

The recurrence of sodium leak incidents across the broader international history of fast reactor operation, documented in detail by International Atomic Energy Agency technical reports cataloguing sodium water and sodium air reaction incidents across multiple national reactor fleets, including a serious 1973 sodium fire at the Soviet Union's BN-350 reactor and a cluster of smaller leaks recorded across the United Kingdom's Prototype Fast Reactor over its operating lifetime, illustrates a pattern directly relevant to this chapter's central argument. Sodium's reactivity with air and water has never been in scientific doubt, not for a single year since Goldschmidt's contemporaries and countless chemistry students before and after him first demonstrated it. What has proven difficult, across seven decades of fast reactor engineering by some of the most sophisticated nuclear engineering organizations in the world, is maintaining, indefinitely and without any single point of failure, the physical integrity of every pipe, weld, sensor housing, and joint in a system built specifically to keep hundreds or thousands of kilograms of that same reactive metal permanently and completely isolated from the atmosphere and from water. Some sodium cooled reactor designs, learning from these very leak incidents, have subsequently incorporated design features specifically intended to reduce this risk, including double walled steam generator tubes at reactors such as the United States' Experimental Breeder Reactor II, a design choice credited with that particular facility operating for thirty years without a single sodium water reaction incident, a genuinely notable engineering achievement that stands as a useful counterpoint to Monju within the same broader technological history, demonstrating that while the underlying chemical hazard cannot be engineered away, it can, with sufficiently rigorous design, be engineered around for extended and genuinely safe periods of operation.

Chapter 4

The Cold Weld: Thermite and the Physics of Concentrated Heat

In 1893, a German chemist named Hans Goldschmidt was working on a problem that had nothing to do with fire, welding, or warfare. He wanted to produce pure metals, free of the carbon contamination that inevitably resulted from the standard industrial method of extracting metal from ore, which relied on heating the ore with carbon to strip away oxygen. The carbon method worked, but it left trace amounts of carbon behind in the finished metal, a contamination that mattered a great deal for certain precision applications. Goldschmidt's insight was to replace carbon with aluminum as the reducing agent, since aluminum has an extraordinarily strong chemical affinity for oxygen, strong enough to pull oxygen away from many other metal oxides entirely. What Goldschmidt discovered, once he began experimenting with mixtures of aluminum powder and iron oxide, was not simply that the reaction worked, but that it worked with a violence he had not fully anticipated: once ignited, the mixture did not merely react, it consumed itself in a self sustaining reaction that produced molten iron and reached temperatures now estimated at somewhere between two thousand and three thousand degrees Celsius, hot enough to melt steel almost instantly. He patented the process in 1897 under the trade name Thermit, and the more generic term thermite has described this entire family of reactions ever since.

The chemistry behind thermite's extraordinary heat output comes down to a single number buried in thermodynamic tables: the enthalpy of formation of aluminum oxide is unusually large and negative, meaning that when aluminum combines with oxygen, it releases an exceptionally large amount of energy, more than almost any other common metal oxidation reaction. In the classic thermite formulation, aluminum powder is mixed with iron oxide, more specifically a form combining iron two oxide and iron three oxide, in roughly a one to three ratio by weight, and when ignited, typically by a magnesium fuse or another high temperature ignition source since the reaction itself requires a substantial activation energy to begin, the aluminum strips oxygen away from the iron, producing aluminum oxide and pure molten iron in a matter of seconds. The reaction is entirely self contained. It requires no external oxygen, since the oxidizer is already present within the iron oxide itself, and this is the single most important practical fact about thermite from a safety perspective: because the reaction supplies its own oxygen, it cannot be smothered, and because it does not depend on atmospheric oxygen at all, water poured onto burning thermite does not extinguish it. Worse, water introduced into a thermite reaction already in progress can flash instantly into steam upon contact with material at two thousand plus degrees Celsius, causing a violent steam explosion that scatters molten metal in every direction, a firefighting hazard that places thermite in the same broad category discussed in the previous chapter, reactions for which water is not a solution but an accelerant of harm.

Goldschmidt's original commercial interest in the reaction was metallurgical rather than destructive, and the application that made thermite an enduring part of industrial infrastructure worldwide was railroad welding. Joining two lengths of rail together had historically relied on mechanical fasteners, bolted joints and fishplates, which worked but left a small gap at every joint, a weak point that caused wear, noise, and the characteristic clacking sound of older rail lines. Thermite welding offered something better: a portable mold placed around the gap between two rail ends, a measured charge of thermite mixture ignited above it, and molten iron flowing directly into the gap, fusing the two rail segments into a single continuous piece of steel once it cooled. The technique required no external power source, no massive equipment, and could be performed essentially anywhere along a rail line, which made it enormously practical for railroad maintenance crews working in remote locations. The process was adopted for welding streetcar track in the German city of Essen within years of Goldschmidt's original patent, and though major railroad companies in some countries were slower to adopt it fully, waiting in some cases until the end of the 1920s, thermite welding eventually became and remains a standard technique for joining rail worldwide, still performed by maintenance crews using essentially the same chemistry Goldschmidt discovered in his laboratory more than a century ago.

The same properties that made thermite valuable for welding, extreme, concentrated, self sustaining heat with no reliance on external oxygen, also made it militarily attractive almost immediately, and this dual identity, essential industrial tool and effective incendiary weapon, has followed thermite through the twentieth century and into the twenty first. During both world wars, thermite formulations were incorporated into incendiary bombs and grenades, valued specifically for their ability to burn through armor plating and ignite structures that conventional explosives alone would not reliably set alight, since thermite's molten iron output could burn its way directly through sheet metal. Thermite based charges have also historically been used in demolition applications, and the same reaction played a documented role in metal purification processes during the Manhattan Project, employed in uranium purification work, an application far removed from either railroad maintenance or warfare but relying on precisely the same underlying chemistry, aluminum's powerful thermodynamic hunger for oxygen.

This dual identity has, in more recent decades, made thermite a recurring subject of public confusion and misinformation, most prominently in connection with conspiracy theories surrounding the collapse of the World Trade Center towers on September 11, 2001, in which some commentators have claimed that thermite or a related compound was used to deliberately demolish the buildings. This book does not exist to adjudicate that claim in detail, which falls well outside its scope and has been extensively examined and rejected by the engineering community and government investigators studying the collapse, including the National Institute of Standards and Technology. What is worth noting here, briefly and for the sake of scientific accuracy rather than political commentary, is a fact about thermite's actual chemical behavior that tends to get lost in that particular controversy: thermite is an intensely effective localized heat source, capable of melting through a small, concentrated area of metal very quickly, but it is a poor tool for producing widespread structural failure across a large steel framed building, precisely because it must be placed and ignited at each specific location where its heat is wanted, it burns relatively slowly compared to a true high explosive, and it does not generate the kind of blast wave or overpressure associated with the sudden, building wide structural failures documented in engineering analyses of the collapse. Readers interested in the science of the September 11 building failures specifically are better served by the detailed structural engineering literature on the subject than by an incendiary chemistry chapter in a broader popular science book, and this chapter mentions the controversy at all only because thermite's genuine industrial and historical importance has, unfortunately, become tangled up in the public mind with a claim its actual combustion chemistry does not straightforwardly support.

Set the conspiracy theories aside, and thermite remains a genuinely instructive case study for the purposes of this book precisely because it represents one extreme end of the exothermic spectrum discussed in the opening chapter: an enormous negative delta H combined with a meaningfully high activation energy, which is exactly the combination that makes thermite simultaneously useful and hazardous in very specific, predictable ways. Because the activation energy is substantial, thermite mixtures are reasonably stable and safe to store and transport when properly formulated and separated from ignition sources, unlike the ammonium nitrate contamination scenarios discussed in a later chapter, where activation energy can be inadvertently and dangerously lowered by external factors. Thermite requires deliberate, sustained ignition, typically well above the temperature a simple spark or open flame can reliably provide, which is precisely why railroad welding crews use a dedicated magnesium ribbon or specialized igniter rather than a match. But once that activation threshold is crossed, the enormous negative delta H takes over completely, and the reaction proceeds to full completion with essentially no way to interrupt it short of physical separation of unreacted material, since it needs no external oxygen and cannot be smothered or drowned in any conventional sense.

The practical safety lessons drawn from a century and more of thermite handling are consequently very specific and very different from the lessons of earlier chapters. Where hay fires are managed through moisture control and temperature monitoring over a period of weeks, and alkali metal hazards are managed by rigorously excluding any water contact, thermite hazards are managed primarily through ignition control and physical containment. Thermite mixtures are stored well away from any spark or flame source, transported in stable, unreacted powder form rather than pre ignited, and used, in every legitimate industrial and welding application, within carefully engineered molds and containment systems designed specifically to direct the molten output where it is wanted and nowhere else. Railroad welding crews wear specialized protective equipment and follow procedures that have themselves been refined over more than a hundred years of practical experience, because the same reaction that makes thermite an elegant, portable, and genuinely brilliant solution to the problem of joining rail in the field remains, without careful handling, capable of producing molten metal at temperatures few other common reactions can match, entirely indifferent to whether the metal it melts belongs to a railroad joint or a bystander's hand.

It is worth walking through the modern procedure in some detail, since the specific steps involved illustrate, in miniature, nearly every safety principle discussed elsewhere in this book, condensed into a working process that a two person rail crew can complete at a remote trackside location in well under an hour. The process begins with rail preparation, cleaning and precisely aligning the two rail ends to be joined and leaving a gap of a carefully specified width, typically in the range of twenty five millimeters, since too narrow a gap prevents proper metal flow while too wide a gap compromises the finished weld's strength. A purpose built ceramic or sand mold is then clamped securely around the joint, sealed at every edge to prevent the molten metal from escaping anywhere except into the intended weld cavity, a containment step directly analogous to the engineered vessels and pressure relief systems examined in this book's chapter on industrial reactor safety, translated here into a single use, field expedient form. Before any thermite is ignited, the rail ends themselves are preheated, typically using a fuel gas torch, to a specified temperature, since pouring molten metal at over two thousand degrees Celsius onto cold steel would cause thermal shock severe enough to crack the rail or compromise the resulting weld's structural integrity, a reminder that even a reaction violent enough to melt iron on contact still requires its surroundings to be properly prepared to receive that energy safely and effectively.

The thermite mixture itself, aluminum and iron oxide powder in carefully measured proportions, along with small quantities of alloying additives used to fine tune the finished weld metal's exact composition, is loaded into a refractory crucible suspended directly above the sealed mold. Ignition is achieved using a specialized high temperature igniting compound, commonly a magnesium based starter mixture applied via a long handled lighter or a purpose made ignition match, specifically because ordinary sparks or open flame, sufficient to ignite far less demanding fuels, cannot reliably deliver the concentrated, sustained heat needed to overcome thermite's substantial activation energy, precisely the same activation energy barrier discussed in the opening chapter of this book that keeps stored thermite powder safely inert through years of transport and handling. Once ignited, industry procedural guides specify that the visible reaction itself typically lasts only eight to fifteen seconds, an almost startlingly brief window during which aluminum strips oxygen from the iron oxide, generating molten iron and a floating layer of aluminum oxide slag, both of which then flow downward from the crucible directly into the sealed mold, filling the gap between the rail ends. Crews are trained to observe this brief window closely, since the visible flame's appearance and disappearance mark, respectively, the beginning and end of the active chemical reaction, information crews use to time every subsequent step in the sequence precisely.

What follows the pour is, in many respects, as procedurally important as the reaction itself. Standard railway welding procedures specify a minimum waiting period, commonly five minutes or more, before the mold and any surrounding fixtures can be safely removed, allowing the newly formed weld to begin solidifying and cooling sufficiently that disturbing it will not compromise the joint's structural integrity, and a further period before the excess metal, the risers and flash that inevitably form around the edges of any casting process, can be safely cut away and the weld surface ground smooth and level with the surrounding rail. Completed welds are then inspected, in modern rail networks, using ultrasonic testing equipment capable of detecting internal defects, porosity, slag inclusions, or hairline cracks, invisible to the naked eye but potentially catastrophic if left undetected in a joint that will subsequently bear the repeated, cyclical stress of heavy rail traffic for years or decades to come, a quality assurance step that connects thermite welding's ancient nineteenth century chemistry directly to the modern nondestructive testing techniques examined more broadly in this book's chapter on detection and prevention science.

None of this procedural elaboration changes the underlying chemistry Goldschmidt discovered in his laboratory more than a century ago. What it demonstrates, instead, is the sheer density of accumulated engineering practice required to take a reaction this powerful, this fast, and this thermally extreme, and make it not merely usable but genuinely, routinely, boringly safe, safe enough that rail maintenance crews around the world perform thousands of these welds every year with a well established, low incident rate, precisely because every step in the modern procedure, preheating, sealed containment, precisely timed demolding, and post weld inspection, has been refined, over decades of accumulated field experience, specifically to work with thermite's genuine hazards rather than to pretend those hazards do not exist. It is, in this sense, one of the more encouraging case studies in this entire book: proof that even one of the most thermally violent common reactions in industrial chemistry can be tamed into routine, reliable use, provided the engineering discipline surrounding it is treated with the same seriousness as the chemistry itself.

Chapter 5

Grain Dust and the Silent Silo

A single failed bearing, roughly the size of a fist, caused the deadliest industrial disaster most Americans have never heard of. On June 8, 1998, at the DeBruce Grain Elevator near Wichita, Kansas, then one of the largest grain storage facilities in the world, a concentrator roller bearing seized after running without adequate lubrication. As the conveyor belt continued rolling over the now frozen roller, friction between the two surfaces drove the temperature at that single point up to approximately 260 degrees Celsius, an effect investigators later described using an old industrial term, the razor strop effect, borrowed from the friction used to sharpen a blade against leather. That temperature was well above the roughly 220 degrees Celsius required to ignite a layer of grain dust, and grain dust, as it happened, was plentiful in exactly that location, having accumulated inside the roller housing over time the way it does in nearly every grain handling facility that does not maintain rigorous housekeeping. What followed was not one explosion but a rapidly escalating series of them, as the initial ignition kicked dust already settled on surfaces throughout the facility up into the air, where it formed new, larger dust clouds that fed subsequent, more powerful blasts, traveling through the elevator's tunnels, galleries, and vertical shafts, propelled in part by the same chimney effect that makes a tall building an efficient conduit for smoke and hot gas. Seven workers died. Ten more were injured. And the mechanism that caused it all, a small amount of dust, a source of heat, and enough confined space to let the resulting pressure build, is common to nearly every serious grain facility explosion documented in more than a century of American agricultural history.

The concept investigators and safety engineers use to describe this mechanism is called the dust explosion pentagon, and it is worth walking through carefully because it explains something that strikes most people as deeply counterintuitive on first encounter: solid grain, the kind found in any farm field or breakfast cereal, does not burn explosively. Ground into fine dust and suspended in air, it can. The explanation lies in surface area. A kernel of wheat or corn presents a relatively small surface area to the surrounding oxygen relative to its total mass, which limits how quickly it can react even if ignited, much the way a log burns slowly in a fireplace. Grind that same kernel into fine dust, and the total surface area exposed to oxygen increases enormously, since the same mass of material is now distributed across millions of tiny particles, each one individually capable of reacting with oxygen far faster than the solid kernel ever could. Suspend that dust in air, rather than leaving it settled on a floor or ledge, and every individual particle gains direct, immediate access to atmospheric oxygen on all sides. The dust explosion pentagon names the five conditions that, together, transform this suspended fine material into an explosive hazard: oxygen, which is simply present in ordinary air; fuel, in the form of combustible dust, and grain dust from wheat, corn, and especially rice and flour is highly combustible; dispersion, meaning the dust must actually be suspended in the air rather than sitting inert on a surface; confinement, an enclosed or partially enclosed space that allows pressure from the initial ignition to build rather than dissipate harmlessly outward; and an ignition source, which can be as small and unremarkable as an overheated bearing, a static discharge, or a spark from metal striking metal.

Remove any single one of these five conditions and a dust explosion cannot occur, which is the central insight behind every modern grain handling safety regulation, and it is also why the story of American grain dust explosions is, in large part, a story about which of the five conditions turned out to be hardest to reliably control at industrial scale. Confinement is essentially unavoidable, since grain elevators are by design tall, largely enclosed structures built to protect stored grain from weather and pests. Oxygen cannot practically be excluded from a working facility where people need to breathe. That leaves fuel, dispersion, and ignition sources as the three points of practical intervention, and the regulatory history of American grain handling over the past half century reflects a sustained, if imperfect, effort to control all three.

The scale of the historical problem is difficult to overstate. Data compiled from a United States Department of Agriculture task force covering the period from 1958 to 1976 documented an average of eight grain elevator explosions annually in the United States, resulting in roughly six deaths and fifteen injuries every year, a steady, grinding toll that persisted for nearly two decades without triggering the kind of sweeping regulatory response the numbers might seem to warrant. What changed public and government attention was a sudden, concentrated spike in catastrophic incidents. In a single one month span from December 1977 into January 1978, five separate grain elevator explosions killed sixty two people and injured fifty three more across the country. Among the worst of these was an explosion at a grain elevator in Westwego, Louisiana, where a buildup of grain dust ignited, blew the top off a silo, and touched off fires in forty eight additional silos at the same facility, ultimately killing thirty six workers, many crushed by collapsing structure rather than burned directly. That same month, an explosion at a Galveston, Texas grain elevator killed twenty workers, an incident investigators later attributed in part to unusually low winter humidity, which had left the stored grain dust drier, and therefore more easily suspended and more readily ignitable, than it would have been in more typical conditions.

This concentrated wave of disaster prompted the federal government to formally engage the National Academy of Sciences in 1978 to convene a symposium specifically examining grain elevator explosions, a direct scientific response to a demonstrated and escalating public safety crisis. It would still take nearly a decade for that scientific engagement to translate into binding regulation. The Occupational Safety and Health Administration's Grain Handling Facilities Standard was not implemented until 1987, nearly ten years after the deadly winter of 1977 into 1978, and the delay itself is a case study in how slowly regulatory response can trail well documented, well understood hazard, even when the underlying chemistry and physics have been understood for decades. Once implemented, the standard's effect was immediate and measurable: grain dust explosions in the United States dropped by an estimated 42 percent in the standard's first year alone, a reduction achieved not through any new scientific discovery but simply through the systematic, mandated application of housekeeping and engineering controls, regular dust removal, restrictions on hot work near accumulated dust, bearing lubrication and maintenance schedules, and dust collection systems designed to capture particulate before it could accumulate to dangerous levels, that the dust explosion pentagon had already identified as the available points of intervention.

The improvement, it is important to note, was not the end of the story, and the DeBruce explosion in 1998, more than a decade after the standard took effect, demonstrates precisely why. Chemical Safety and Hazard Investigation Board data covering 1980 through 2006 documented an average of eleven grain dust explosions per year in the United States even after the OSHA standard's implementation, with an average of four and a half fatalities annually, a substantial improvement over the pre regulation era but nowhere close to elimination of the hazard. A subsequent CSB reporting period covering 2006 through 2017 found nine explosions and five fatalities per year on average, figures that suggest the underlying hazard has been meaningfully reduced but not resolved. Purdue University's Department of Agricultural and Biological Engineering, which has tracked grain dust incidents through an annual report for years, recorded seven reported grain dust explosions across United States food and agricultural facilities in 2017 alone, resulting in five fatalities and twelve injuries, figures that remained, at the time, below the prior decade's average of 9.3 explosions per year but still represented real workplace deaths occurring in an industry with well established, well understood, and largely preventable hazard controls.

The persistence of the problem, even after decades of regulation and a clear scientific understanding of its cause, illustrates something this book returns to repeatedly: understanding a hazard and eliminating it are two very different achievements, and the gap between them is usually filled by maintenance schedules, budget decisions, staffing levels, and the countless small daily choices about how thoroughly to clean a facility that no regulation, however well written, can fully guarantee will be followed on every shift, in every facility, every single day. The DeBruce bearing that seized in June 1998 had, according to investigators, been running without adequate lubrication, a maintenance failure that had nothing to do with any gap in scientific understanding and everything to do with the ordinary, unglamorous discipline of routine facility upkeep. The dust that fed the resulting explosions had accumulated in locations throughout the facility over time, in exactly the pattern that OSHA's 1987 standard was specifically designed to prevent through mandated housekeeping schedules.

This is, in a sense, the quiet tragedy embedded in the statistics of grain dust explosions more than in any single catastrophic event: the hazard is not mysterious, has not been mysterious for well over a century, and the specific engineering and procedural controls needed to prevent it are neither exotic nor prohibitively expensive relative to the value of the facilities and lives involved. What grain elevator explosions require, more than any breakthrough in chemistry or engineering, is sustained attention to an invisible hazard that, on any ordinary working day, looks like nothing more than a faint haze of dust in the air and a light coating of particulate on overhead surfaces, the kind of unremarkable industrial grime that is easy to notice and easier still to defer cleaning until tomorrow. The chemistry does not defer. It waits, patiently, in whatever quantity accumulates, for the specific and entirely ordinary combination of a heat source and a moment of dispersion that turns settled dust into an explosive cloud, and it has been waiting, in grain facilities across the world, for as long as humans have been storing and processing large volumes of dry organic material in enclosed spaces.

A decade after the OSHA Grain Handling Facilities Standard took effect, a strikingly similar disaster demonstrated that the same underlying hazard extends well beyond grain in the strictest sense, to any sufficiently fine, sufficiently combustible organic dust handled at industrial scale. On the evening of February 7, 2008, at the Imperial Sugar refinery in Port Wentworth, Georgia, a sugar dust explosion occurred inside an enclosed steel conveyor belt located directly beneath two large sugar storage silos. The explosion lofted years of accumulated sugar dust that had settled on floors and elevated horizontal surfaces throughout the facility's packing buildings, triggering a rapid cascade of secondary explosions that tore through the plant with a violence the CSB's subsequent investigation described in vivid physical detail: pressure waves powerful enough to heave thick concrete floors upward and collapse brick walls, in the process blocking stairwells and other exit routes that workers needed to escape the resulting fires. Fourteen workers died. Thirty six more were treated for serious burns and injuries, some resulting in permanent, life altering conditions. The explosions and subsequent fires ultimately destroyed the facility's sugar packing buildings, its palletizer room, and its storage silos entirely, while severely damaging the bulk train car loading area and portions of the refining process itself.

What makes the Imperial Sugar disaster particularly instructive, and particularly sobering, for the purposes of this book is the specific finding that emerged from the CSB's investigation regarding how long the underlying hazard had been known to the company's own management before it finally erupted. Internal company records and material safety data sheets, according to subsequent independent review of the investigation's findings, documented awareness of combustible sugar dust hazards dating back to at least 1925, more than eight decades before the explosion, including documented data indicating that sugar dust could become explosive at concentrations as low as roughly four hundredths of a gram per liter of air, an exceedingly small quantity, easily achieved by ordinary operational dust release, let alone by dust freshly lofted from thick accumulations that CSB investigators later found had built up to depths of several inches in some areas of the facility. This was not, in other words, a hazard the company discovered only in the investigation's aftermath. It was a hazard the company had documented, in its own internal safety literature, for the better part of a century, and had nonetheless allowed to accumulate, inspection after inspection, shift after shift, to a degree that made a catastrophic secondary explosion cascade all but inevitable once any ignition source, ultimately traced to an overheated conveyor bearing operating beneath sugar handling equipment that had itself been inadequately designed and maintained to minimize dust release into the surrounding work area, eventually appeared.

The CSB's investigation identified a further contributing factor that adds a genuinely unsettling twist to the Imperial Sugar case, one directly relevant to the engineering safeguards discussed throughout this book: steel cover panels had been installed on the conveyor belt where the initial explosion occurred less than a year before the disaster, a modification investigators concluded had likely been intended, at least in part, to control dust release from the conveyor system, but which instead created an enclosed space in which airborne sugar dust could accumulate above the minimum explosible concentration far more readily than it could have in the conveyor's previous, more open configuration, inadvertently satisfying the confinement condition of the dust explosion pentagon described earlier in this chapter in the course of an engineering change intended to improve, not worsen, the facility's dust management. This detail deserves particular attention because it illustrates a genuinely difficult and recurring problem in industrial safety engineering more broadly: a modification made with good intentions, evaluated without a full and rigorous understanding of how it interacts with a hazard the modifying engineers may not have fully appreciated, can sometimes make a known danger measurably worse rather than better, a possibility that underscores why modern process safety management increasingly emphasizes formal hazard review procedures for even seemingly minor equipment changes, rather than relying on the informal judgment of whoever happens to be implementing a given modification at the time.

The regulatory aftermath of Imperial Sugar followed a now familiar pattern from this chapter's earlier account of the 1977 and 1978 wave of grain elevator explosions that preceded the original OSHA standard, though with a notably less complete resolution. Within months of the disaster, OSHA announced its intention to begin developing a comprehensive combustible dust standard covering industries beyond grain handling specifically, and the CSB's final report, released in 2009, strongly reiterated that recommendation. As of assessments conducted roughly a decade after the disaster, however, comprehensive federal rule making on combustible dust hazards across general industry had still not been finalized, a regulatory gap that safety researchers and industry observers have continued to highlight, and one that stands as a sober reminder of a pattern this book has traced across multiple chapters and multiple decades: the lag between a well documented, thoroughly investigated hazard and the completion of binding regulation intended to address it comprehensively can, in practice, extend for years or even decades beyond the disaster that most clearly demonstrated the hazard's real world consequences.

Chapter 6

A Wider Family: Metal Dust and the Kunshan Disaster

The dust explosion pentagon introduced in an earlier chapter of this book, oxygen, fuel, dispersion, confinement, and an ignition source, does not care what the fuel is made of. Grain dust satisfies all five conditions readily enough to have killed hundreds of American farm and elevator workers across the twentieth century, but grain is, relative to what this chapter is about, a comparatively mild example of a combustible dust hazard. A far more violent and far less widely appreciated category of dust explosion involves not organic material at all, but finely divided metal, and the worst single incident in the recorded history of this specific hazard occurred not in an American grain belt but inside an automotive parts factory in eastern China, on a Saturday morning in August 2014, in a workshop where laborers were doing nothing more unusual than polishing wheels.

The Kunshan Zhongrong Metal Products Company, located in Kunshan, in China's Jiangsu province, operated a factory that produced aluminum alloy wheel hubs for a client list that included General Motors, among the largest and most recognizable automotive manufacturers in the world. Part of the finishing process for these wheels involved polishing their surfaces to achieve the smooth, bright metal finish automotive customers expect, a process that, like most metal grinding and polishing operations, generates enormously fine metal particulate as an unavoidable byproduct, aluminum dust so fine that individual particles are measured in micrometers, small enough to remain suspended in air for extended periods rather than settling quickly to the floor the way coarser dust would. At approximately 7:37 in the morning on August 2, 2014, with more than two hundred sixty workers present in the facility, a higher than usual number that day because weekend overtime pay had drawn additional staff in, a dust explosion tore through the workshop where the polishing was taking place. Seventy five workers were killed at or near the scene. A further one hundred eighty five were injured, many of them severely, and in the weeks and months that followed, seventy one of the seriously injured ultimately succumbed to their wounds as well, bringing the disaster's eventual total death toll to one hundred forty six, a figure that establishes the Kunshan explosion as one of the deadliest dust explosions ever documented anywhere in the world, in any industry, exceeded in scale by only a handful of historical coal mine dust explosions.

The specific mechanism behind the disaster's severity connects directly to a chemical property of aluminum this book has already examined from a very different angle in an earlier chapter: aluminum's exceptionally strong affinity for oxygen, the same underlying thermodynamic property responsible for thermite's ability to melt steel on contact. In bulk, solid form, aluminum is entirely unremarkable and thoroughly safe, protected from ready oxidation by a thin, naturally forming oxide layer on its surface that most people encounter constantly in aluminum foil, cans, and countless manufactured products without a second thought. Ground into sufficiently fine dust, however, that protective oxide layer becomes proportionally almost irrelevant, since the ratio of fresh, reactive metal surface area to total particle volume increases dramatically as particle size decreases, precisely the same surface area principle discussed in this book's earlier chapter on grain dust, but combined here with a base material considerably more energetic, gram for gram, than any organic dust could ever be. Investigators who subsequently studied the Kunshan disaster in detail, publishing their findings in the peer reviewed engineering literature, concluded that a series of consecutive explosions had originated within one of the facility's external dust collection filters, precisely the equipment installed, ironically, to capture and remove the very dust that was accumulating instead in dangerous quantities within the collection system itself, a system investigators found had not been adequately designed, ventilated, or maintained to manage the volume and fineness of aluminum dust the polishing operation was actually generating.

Chinese government investigators who examined the disaster's underlying causes reached conclusions that will, by this point in the book, sound distinctly and unhappily familiar to readers who have made their way through the earlier chapters on grain elevators and industrial reactors. The investigation found that Kunshan Zhongrong had failed to properly store dangerous materials, lacked appropriate ventilation and dust removal systems adequate to the hazard its own polishing operations were generating, and operated with a substandard electrical system that likely provided the ignition source, whether a spark from faulty wiring or an overheated piece of equipment, that ultimately touched off the explosion. Investigators further found that the company had received prior warnings specifically about elevated dust levels in the facility, according to reporting citing the local workplace safety authority, warnings that evidently went unaddressed in any way sufficient to prevent the disaster that followed, a pattern of documented prior warning ignored until catastrophic consequence this book has now traced across grain elevators, a fertilizer warehouse, and, in this chapter, an automotive parts factory on the other side of the world. A senior Chinese work safety official, in remarks reported by the state news agency Xinhua in the disaster's immediate aftermath, characterized the failures that led to the explosion as amounting to very serious dereliction of duty, a blunt official assessment that echoes, in less technical language, the same conclusions that formal investigative bodies like the United States Chemical Safety and Hazard Investigation Board have reached, repeatedly and across multiple decades, about American industrial disasters examined elsewhere in this book.

The Kunshan disaster is far from an isolated case within the specific category of metal dust hazards, and a broader review of documented incidents makes clear that aluminum is only one entry, if a particularly prominent one, in a wider family of combustible metal dusts that includes magnesium, titanium, and zirconium, each of which shares the same underlying vulnerability: a fine particle form dramatically more reactive than the same metal in bulk, solid form, and each of which has produced its own documented history of industrial explosions. A 2009 explosion involving aluminum powder in a factory being used as temporary worker housing in the Chinese city of Danyang killed eleven people and injured twenty more. A 2012 explosion in an aluminum lock polishing workshop in the city of Wenzhou killed thirteen and injured fifteen. Titanium dust, valued in fireworks and pyrotechnics specifically for the same extreme reactivity that makes it hazardous in an industrial setting, was implicated in a fatal explosion at an A.L. Solutions plant in New Cumberland, West Virginia, that killed three workers, an incident that helped draw broader American regulatory and industry attention to metal dust hazards that had, for years prior, received considerably less systematic attention than organic dust hazards like grain despite involving, gram for gram, an often more energetic and more difficult to suppress category of explosive risk.

A particularly instructive body of evidence on this point comes from a detailed incident history compiled by AMG Alpoco, an aluminum powder manufacturing facility in Anglesey, Wales, which has operated since the late 1960s and which published a candid, self examining account of six separate aluminum dust explosion incidents that occurred at its own site across five decades of operation. The company's own retrospective analysis reached a conclusion with genuinely broad implications for how this book's readers should think about combustible metal dust as a category: that the explosion prevention and protection practices considered adequate by the engineering standards of any given era frequently proved, in the plant's own accumulated experience, insufficient in practice, and that reliably managing fine aluminum powder's extreme reactivity and rapid flame propagation characteristics has repeatedly pushed at or beyond the limits of what the best available protective technology of the time could handle with full confidence. This is a notably more humble and more scientifically honest assessment than the confident, case closed tone that sometimes characterizes safety literature written after a single incident, and it is included here specifically because it complicates any simple narrative in which better engineering alone straightforwardly and permanently solves a hazard of this kind. AMG Alpoco's response to this hard won institutional experience, according to its own published account, has ultimately involved restricting personnel access even to plant and equipment specifically designed and certified for explosion protection, treating the fundamental unpredictability of fine aluminum powder's behavior as a permanent operating constraint to be continuously managed rather than a problem that any single generation of engineering improvement could be expected to fully and finally resolve.

This chapter's inclusion in a book organized around exothermic chemistry connects back most directly to the concept of surface area and reaction rate introduced in this book's opening chapter, and it is worth drawing that connection explicitly as this chapter closes. A solid ingot of aluminum, a solid block of magnesium, a solid rod of titanium, all present real but manageable fire risk under most industrial conditions, since their comparatively low surface area to volume ratio limits how quickly they can react with atmospheric oxygen even when directly exposed to flame. Reduce any of these same metals to a fine enough particle size, disperse that particle cloud into the air of an enclosed workshop, and introduce a single adequate ignition source, and the identical underlying chemistry, aluminum's or magnesium's or titanium's fundamental thermodynamic hunger for oxygen, expresses itself instead as an explosion capable, as the Kunshan disaster demonstrated with terrible clarity, of killing well over a hundred people within a single industrial facility in the space of a few catastrophic seconds. The metal itself never changed. What changed, as in nearly every case study examined throughout this book, was the physical form in which that metal's underlying chemistry was allowed to express itself, and whether the people responsible for the facility where that transformation was occurring had taken seriously enough the well documented, decades old engineering literature describing exactly how and why it happens.

There is a further complication specific to metal dust fires that deserves mention here, since it connects this chapter directly back to the discussion of water as an inappropriate suppression agent examined in this book's earlier chapter on reactive metals and water. Firefighters responding to a metal dust fire, particularly one involving aluminum, magnesium, or titanium, face a genuinely counterintuitive hazard beyond the initial explosion risk itself, since these same metals, in sufficiently fine or molten form, can react vigorously and dangerously with water in a manner directly analogous to the alkali metal reactions examined earlier in this book, generating hydrogen gas and additional heat that can worsen rather than suppress an ongoing fire. Standard firefighting guidance for metal fires accordingly designates them as a distinct hazard class, Class D fires under the American fire classification system, requiring specialized dry powder extinguishing agents specifically formulated to smother the burning metal without introducing water or, in many cases, without introducing the carbon dioxide or standard dry chemical agents that work well against ordinary combustible or electrical fires but that can, in the case of certain burning metals, actually be consumed as a reactant by the burning metal rather than serving to extinguish it, a metal fire in the most extreme documented cases capable of stripping oxygen directly from carbon dioxide molecules, continuing to burn even when deprived of atmospheric oxygen entirely. This specialized suppression requirement means that fire departments responding to an active industrial fire without advance knowledge that combustible metal dust or material is present risk making the situation meaningfully worse through an entirely reasonable, well trained response calibrated to a more ordinary structural fire, one further illustration of why accurate, specific, and readily accessible hazard information, the kind formal process safety documentation and facility hazard placarding are specifically designed to provide, matters as much in the moments immediately following an ignition event as it does in the years of engineering and procedural work meant to prevent that ignition from ever occurring in the first place.

Chapter 7

The Fertilizer Paradox: Ammonium Nitrate and the West, Texas Disaster

At approximately 7:29 in the evening on April 17, 2013, residents of the small town of West, Texas, population roughly 2,800, began reporting smoke and flames at the West Fertilizer Company's storage and distribution facility, located along Interstate 35 not far from Waco. Volunteer firefighters responded within minutes, and a police officer who arrived early recognized the severity of what he was seeing quickly enough to begin evacuating a nearby basketball court and playground before sealing off traffic access to the area, a decision that almost certainly saved lives. It was not enough to save everyone. Roughly twenty minutes after the fire was first reported, the flames reached approximately thirty tons of ammonium nitrate fertilizer stored inside a wooden receiving bin within the facility, and the resulting detonation killed fifteen people, twelve of them emergency responders who were actively fighting the fire when the building they were standing near ceased, in the span of a fraction of a second, to be a building at all. More than two hundred and sixty others were injured. The blast registered on seismographs as a magnitude 2.1 earthquake and was powerful enough to be detected and analyzed using standard geological monitoring equipment, an unusual and telling detail for what was, at its root, an agricultural fertilizer accident rather than a military or mining event.

To understand why a substance sold openly to farmers across the world, used by the ton on ordinary fields without incident for the overwhelming majority of its history, is capable of producing an explosion powerful enough to register as an earthquake requires returning to the concept introduced early in this book: the distinction between the total energy a reaction can release and the activation energy required to release it. Ammonium nitrate is, under normal storage conditions, chemically stable. It does not spontaneously detonate sitting in a warehouse, a fact confirmed by the sheer scale at which it is produced, stored, and used worldwide as one of the most common nitrogen fertilizers in industrial agriculture, without incident, every single day. But ammonium nitrate's stability is conditional rather than absolute, and the conditions that erode it are, unfortunately, precisely the conditions that can arise during an uncontrolled structural fire: high sustained temperature, confinement that allows pressure and heat to build rather than dissipate, and contamination with combustible material, all three of which were present at the West Fertilizer facility on the night of the explosion, where the wooden storage bin itself, combined with significant quantities of combustible seed stored in the same building, provided exactly the fuel and confinement that pure ammonium nitrate under open, unconfined conditions would have lacked.

This is not a novel or recently discovered vulnerability. It is, in fact, one of the oldest and most thoroughly documented hazards in industrial chemistry, and the West, Texas disaster occupies a place in a lineage of remarkably similar catastrophes stretching back nearly a century. On September 21, 1921, at a BASF chemical plant in the town of Oppau, Germany, workers faced a practical problem: roughly 4,500 tons of a fertilizer mixture combining ammonium nitrate and ammonium sulfate had solidified into a dense, compacted mass inside a large storage silo, and needed to be broken apart for handling and shipment. Their solution, based on the sincerely held and, at the time, widely shared belief that ammonium nitrate mixed with ammonium sulfate in roughly equal proportions was non explosive, a belief with some genuine chemical basis since pure ammonium sulfate dilutes and stabilizes the mixture under most conditions, was to use small dynamite charges to loosen the compacted material, a technique that had reportedly been used at the same facility many times before without incident. On this occasion, it detonated the entire stockpile. The resulting blast carved a crater ninety meters wide and nineteen meters deep into the earth, destroyed approximately eighty percent of the housing in the town of Oppau, tore roofs off buildings as far as twenty five kilometers away, and killed, by most historical accounts, somewhere between five hundred and six hundred people, making it one of the deadliest industrial accidents in European history. The uncomfortable lesson embedded in Oppau is that the workers were not reckless by the standards of their own understanding. They believed, based on prior experience and prevailing chemical knowledge, that the specific mixture they were handling was safe from detonation. They were wrong, and the mechanism by which they were wrong, namely that the mixture's actual composition and moisture content on that particular day differed from the assumptions underlying their belief in its safety, remains a documented and cautionary example of how dangerous it is to generalize chemical safety from limited prior experience rather than from a full understanding of the underlying reaction chemistry.

A little over a quarter century later, on the morning of April 16, 1947, the same underlying chemistry produced what remains the deadliest industrial accident in United States history. The French registered cargo ship Grandcamp, docked in the port of Texas City, Texas, on Galveston Bay, was in the process of being loaded with approximately 2,300 tons of ammonium nitrate fertilizer, cargo intended to aid struggling farmers in a Europe still recovering from the devastation of the Second World War, when a fire broke out in the ship's cargo hold. The ship's captain, reportedly concerned about damaging the valuable cargo, made the fateful decision to fight the fire using the vessel's steam system rather than flooding the hold with water, a choice that, in light of everything discussed in this book about the relationship between heat, confinement, and exothermic decomposition, may have inadvertently done the opposite of what was intended, sealing the burning cargo into an increasingly hot and pressurized confined space rather than cooling it. At approximately 9:12 in the morning, the Grandcamp's cargo detonated. The blast killed at least 581 people, injured approximately 5,000 more, shattered windows as far as twenty five miles from the port, and threw large pieces of steel debris more than a mile from the dock, including the ship's own anchor, which investigators later found nearly two miles from the site of the explosion. The disaster did not end there. The force of the initial blast ignited fires aboard a second nearby ship, the High Flyer, which was also carrying a cargo of ammonium nitrate, and this second vessel exploded later the same day, compounding the destruction across the port.

Placed side by side, Oppau in 1921, Texas City in 1947, and West, Texas in 2013, span nearly a century and represent three entirely distinct national contexts, industrial settings, and immediate causes, a dynamite charge intended to break up compacted fertilizer, a shipboard fire fought with steam rather than water, and an uncontrolled structural fire in a wooden storage building. And yet the underlying chemical mechanism connecting all three incidents is identical: ammonium nitrate, ordinarily stable under normal storage and handling conditions, encountering a combination of sustained heat, confinement, and in most cases contamination with combustible material, conditions that together lower the effective activation energy required to trigger the compound's violent exothermic decomposition, at which point a fertilizer that had, in every one of these cases, been handled routinely and without incident for years prior, converted with terrifying speed into a detonation on a scale associated more commonly with military ordnance than agricultural commerce. This repetition across a century, involving different countries, different regulatory regimes, and different specific triggering circumstances, is precisely the pattern that safety researchers point to when they describe the chemical industry's persistent difficulty translating hard won lessons from one disaster into durable, lasting institutional practice across the wider industry. A safety analysis of the West Fertilizer incident, prepared in the aftermath of the 2013 disaster, noted with evident frustration that despite substantial experience with ammonium nitrate accidents over the preceding century, global learning from that experience had remained poor, a conclusion difficult to dispute given that the specific hazard, fire reaching stored ammonium nitrate in a confined space, had already produced two of the deadliest industrial accidents in modern history before it happened a third time in a small Texas town in 2013.

The United States Chemical Safety and Hazard Investigation Board's final report on the West Fertilizer explosion, issued in January 2016 after an extensive multi year investigation, concluded that the construction of the storage bins themselves, built from combustible wooden materials rather than fire resistant construction, along with the facility's lack of an automatic sprinkler system, plausibly contributed to the severity and scale of the resulting explosion. Investigators explored several specific scenarios for how the fire ultimately triggered detonation rather than simply burning itself out, including the possibility that the ammonium nitrate had become contaminated with carbonaceous soot from the fire itself, since even a small percentage of carbon contamination, as little as two tenths of one percent by some estimates, measurably increases the material's blast sensitivity and detonation potential. Separately, a criminal investigation led jointly by the Bureau of Alcohol, Tobacco, Firearms and Explosives and the Texas State Fire Marshal's office spent years investigating the origin of the initial fire itself, and in 2016 authorities announced that they had concluded the fire had been intentionally set, ruling out accidental and natural causes after conducting more than four hundred interviews, though no arrests were ever made in connection with that determination. The question of who or what specifically ignited the original fire at West Fertilizer, in other words, remains formally unresolved even as the chemistry of what happened after that fire reached the stored ammonium nitrate is understood in exhaustive, well documented detail.

Seven years after West, Texas, and almost exactly a century after Oppau, the same underlying chemistry produced a catastrophe on a scale that dwarfed all of the disasters examined so far in this chapter, and that has since become, alongside Chernobyl and Bhopal, one of the most widely recognized industrial disasters of the twenty first century. On the afternoon of August 4, 2020, a fire broke out in a warehouse at the Port of Beirut, in Lebanon, and spread to an adjacent warehouse, designated Hangar 12, where approximately 2,750 tons of ammonium nitrate had been stored, largely without any dedicated safety measures, for nearly six years. At 6:08 in the evening local time, that stockpile detonated in an explosion that subsequent analysis places among the largest non nuclear explosions ever recorded, producing a crater roughly 124 meters in diameter, registering as the equivalent of a magnitude 3.3 earthquake, and generating a blast wave and heat pulse that killed more than two hundred people, injured over seven thousand more, left an estimated three hundred thousand residents of Beirut homeless, and caused property damage subsequently estimated at more than fifteen billion dollars, devastating a significant portion of one of the Middle East's major cities in the span of a few catastrophic seconds.

The specific history of how 2,750 tons of ammonium nitrate came to be sitting, essentially unmanaged, in a Beirut port warehouse for nearly six years reads as a case study in bureaucratic failure every bit as instructive, in its own way, as the chemistry of the detonation itself. The cargo had originally been loaded aboard a Moldovan flagged vessel, the Rhosus, in 2013, bound for Mozambique, but the ship was detained at the Port of Beirut over unrelated documentation and seaworthiness issues, and its owner subsequently abandoned both the vessel and its cargo rather than resolve the dispute. Lebanese port authorities, left holding a cargo nobody wanted and with nowhere else obviously appropriate to put it, offloaded the ammonium nitrate into the port's Hangar 12 in October 2014, where it remained for the following six years while, according to subsequent investigative reporting by Human Rights Watch and other organizations, Lebanese customs officials repeatedly petitioned various government ministries and courts, across multiple separate written requests over several years, for guidance on what to do with the material, requests that went largely unanswered or unresolved by an administrative and judicial system investigators later characterized as chronically dysfunctional. Reporting following the disaster indicated that port and customs officials had, at multiple points during those six years, specifically flagged the ammonium nitrate as a serious safety hazard requiring urgent resolution, with at least one inspection team reportedly warning, in terms that proved chillingly accurate, that the material was dangerous enough to, in essence, destroy a significant part of the city if it were ever to ignite.

Subsequent technical investigation into the disaster identified several specific factors that combined to make the Beirut explosion considerably more severe than a comparable quantity of properly stored, properly separated ammonium nitrate might have produced. Investigators working with the research and advocacy organization Forensic Architecture, drawing on photographic and video evidence of the warehouse's interior in the months before the explosion, found that the ammonium nitrate had reportedly been stored in violation of widely recognized international safety standards, including guidance published by regulatory bodies such as the United Kingdom's Health and Safety Executive, which specifically prohibit storing combustible or explosive materials in close proximity to bulk ammonium nitrate. According to that same investigative work, the same warehouse also reportedly contained a substantial quantity of fireworks, along with roughly a thousand car tires, both materials whose presence directly alongside the ammonium nitrate stockpile would have provided exactly the kind of combustible fuel and confinement conducive to detonation that this chapter has already traced through the West, Texas and Oppau disasters, a repetition of essentially the identical contributing hazard, ammonium nitrate stored in direct proximity to combustible material inside an enclosed structure, across three catastrophes spanning very nearly a full century.

The Beirut disaster's inclusion at the end of this chapter is meant to underscore, with maximum possible weight, the argument this chapter has built across Oppau, Texas City, and West: the chemistry of ammonium nitrate decomposition has been established science since long before any of these four disasters occurred, and every one of them, separated by decades and by entirely different regulatory, political, and institutional contexts, ultimately traces back to the same recurring failure, a large quantity of ammonium nitrate, stored without the specific separation, containment, and fire prevention measures the known chemistry demands, left to sit for an extended period until an ignition source, whether a demolition charge, a shipboard fire, an unattended structural fire, or a warehouse blaze that began in an entirely separate part of the same port, finally arrived. Beirut adds a particularly stark coda to that pattern because the danger, in this instance, was not merely theoretical or retrospectively obvious. It had been explicitly, repeatedly, and specifically flagged by the officials responsible for the material years in advance, in writing, through official channels, and still nothing was done before the warning proved catastrophically correct.

In the aftermath of the West, Texas disaster, regulatory attention in the United States turned once again, as it had after Oppau and Texas City before it, toward strengthening storage and handling requirements for ammonium nitrate, an effort informed by the specific findings of the CSB investigation regarding building construction materials, sprinkler systems, and safe separation distances between fertilizer storage and surrounding residential development, since one of the most striking features of the West disaster, as with several of its historical predecessors, was the close proximity of the storage facility to homes, a school, and a nursing home, all of which sustained significant damage or destruction in the blast. Whether these regulatory responses will prove more durable and more thoroughly internalized across the fertilizer storage industry than the lessons drawn from Oppau in 1921 or Texas City in 1947 is not a question this book can answer with confidence, since the honest and somewhat unsettling truth revealed by placing these three disasters side by side is that the chemistry involved has never been the limiting factor. The chemistry of ammonium nitrate decomposition has been understood, in essentially its current form, for the better part of a century. What has proven far harder to sustain, across changing generations of workers, facility owners, and regulators, each of whom must independently absorb and act on lessons that a previous generation learned at devastating cost, is the institutional memory required to ensure that understanding a hazard translates reliably into never repeating it.

The specific chemistry underlying ammonium nitrate's transformation from a stable fertilizer into a detonating hazard deserves a closer look, since it illustrates the interplay between decomposition pathway and confinement in a particularly clear and instructive way. Under gradual, low temperature heating in an open, unconfined setting, ammonium nitrate decomposes relatively gently, releasing nitrous oxide and water vapor in a reaction that, while still exothermic, proceeds too slowly and too incompletely to generate the sudden overpressure associated with a detonation. This is, in fact, the reaction responsible for one of ammonium nitrate's own legitimate industrial uses as an oxidizer in certain commercial blasting formulations, where its decomposition is deliberately and carefully controlled rather than allowed to run away. Under the specific combination of high sustained temperature, confinement sufficient to prevent decomposition gases from venting freely, and particularly the presence of even small quantities of combustible contamination, fuel oil, coal dust, or in the case of West, Texas, the carbonaceous soot generated by the fire itself along with combustible seed stored in close proximity, the decomposition pathway shifts toward a considerably more violent and complete breakdown, producing nitrogen, oxygen, and water vapor in a reaction that proceeds at detonation velocity rather than as a comparatively gentle thermal decomposition, releasing its full stored chemical energy within a fraction of a second rather than over an extended period. This sensitivity to contamination explains why formulations combining ammonium nitrate directly with fuel oil, a mixture commonly abbreviated ANFO, have long been deliberately manufactured and used as a commercial and mining explosive precisely because that specific combination reliably produces detonation under controlled initiation, the very same chemical vulnerability that, when it occurs by accident rather than by design, as it did at West, Texas, produces catastrophe rather than controlled demolition.

This dual identity, the same fundamental chemistry serving simultaneously as one of the world's most widely used nitrogen fertilizers and as a component of one of the world's most widely used commercial explosives, has inevitably attracted regulatory and security attention well beyond ordinary industrial safety concerns, particularly following incidents in which ammonium nitrate has been deliberately used as a weapon rather than accidentally detonated through fire or contamination, a distinction that lies outside the primary scope of this book but that has meaningfully shaped the regulatory environment surrounding the fertilizer's sale, storage, and transport in the decades since. In the United States, this has translated into reporting and security requirements administered by the Department of Homeland Security for facilities storing ammonium nitrate above specified quantity thresholds, layered on top of the more purely safety oriented storage and construction standards that emerged from the CSB's investigation into the accidental fire and detonation hazard specifically. The result is a regulatory landscape considerably more complex than existed at the time of the Oppau disaster in 1921, when the very concept of ammonium nitrate as a detonation hazard was still being actively debated among the chemists and engineers responsible for the Oppau facility's operation, several of whom, as noted earlier in this chapter, sincerely believed their specific stockpiled mixture to be non explosive right up until the moment it proved otherwise.

Chapter 8

Runaway in the Reactor: Industrial Batch Chemistry Gone Wrong

At 1:23 in the afternoon on December 19, 2007, a process operator at T2 Laboratories in Jacksonville, Florida, noticed a cooling problem developing in a batch reactor and asked an outside operator to call the company's owners and request that they return to the site. This was the company's one hundred seventy fifth batch of methylcyclopentadienyl manganese tricarbonyl, an antiknock fuel additive commonly abbreviated MCMT, produced using a process the twelve person company had run successfully many times before. One of the two owners arrived and went directly to the control room to help assess the situation. Ten minutes later, at 1:33, the reactor burst. Its contents exploded with force the United States Chemical Safety and Hazard Investigation Board later calculated as equivalent to roughly fourteen hundred pounds of TNT, an explosion powerful enough to be felt and heard fifteen miles away in downtown Jacksonville. The blast killed the owner and the process operator, both of whom had been in the control room, along with two outside operators who were in the process of leaving the reactor area when the vessel failed. Twenty eight additional people, working in businesses near but not part of the T2 facility, were injured by walls and windows that collapsed inward from buildings a quarter mile from the explosion's source. Debris was later found as far as a mile away.

The CSB's subsequent investigation, one of the most extensive the agency has conducted, determined that a runaway exothermic reaction had occurred during the first stage of the MCMT manufacturing process, known as the metalation step, and that the most probable cause was a loss of adequate cooling that allowed pressure and temperature inside the sealed reactor to climb without any effective mechanism to release them, until the vessel itself gave way. This finding, on its own, might read as a fairly ordinary industrial accident report, the kind of dry, technical conclusion that appears in safety literature without necessarily illuminating why the disaster happened at all. What makes the T2 case so significant, and so directly relevant to the themes of this book, is a detail buried a little deeper in the investigation: the CSB found that although both owners of the company held undergraduate degrees, one in chemistry and one in chemical engineering, neither of them, nor apparently anyone else working at the facility, fully appreciated the potential consequences of a runaway chemical reaction in the specific process they were operating. This was not a case of untrained workers cutting corners in an industry they barely understood. It was a case of formally educated chemists and engineers running a process for years, one hundred seventy four successful batches deep, without recognizing that the underlying reaction chemistry carried a latent thermal runaway hazard that a sufficiently severe cooling failure could turn catastrophic in a matter of minutes.

The investigation's broader finding, and the one that generated the most discussion within the chemical engineering profession afterward, was that this gap in understanding was not unique to the two T2 owners but reflected a genuine, structural gap in how chemical engineers are trained. The CSB's report noted that most baccalaureate chemical engineering curricula in the United States did not, at the time, specifically address reactive hazard recognition or management as a distinct area of study, with less than eleven percent of surveyed university programs including dedicated coursework on the subject. A chemical engineer could, and evidently often did, complete a full undergraduate education and enter industrial practice without ever formally studying the specific warning signs, the specific calorimetric testing methods, or the specific engineering safeguards associated with reactions capable of accelerating beyond a facility's ability to control them. The CSB's central recommendation following the T2 investigation was not a call for new regulation in the traditional sense, but a call directed at the American Institute of Chemical Engineers and the Accreditation Board for Engineering and Technology to work together to close this specific educational gap, on the theory that an engineer who has never been taught to recognize a reactive hazard cannot reasonably be expected to design around one.

This educational blind spot connects directly to the thermal runaway mechanism introduced in the first chapter of this book, the exponential relationship between temperature and reaction rate first formalized by Svante Arrhenius in the 1880s, more than a century before four people died at T2 Laboratories. The chemistry that killed them was neither new nor obscure. Arrhenius kinetics have been a standard part of physical chemistry curricula for generations. What was evidently missing, at T2 and, according to the CSB's broader survey, at a meaningful fraction of American chemical engineering programs more generally, was the specific translation of that abstract kinetic principle into the applied, practical judgment required to recognize when a particular industrial process, however routine it has become through repetition, still carries the latent capacity for a self accelerating feedback loop between heat and reaction rate that a cooling system failure can trigger within minutes. The one hundred seventy four successful prior batches at T2, rather than being evidence of safety, were in a stricter sense evidence of nothing at all regarding what would happen on the one hundred seventy fifth run if cooling failed at a critical moment, since a system running safely under adequate cooling reveals very little about how that same system behaves once cooling is compromised, a distinction the CSB report suggests the company's chemists had not fully internalized.

A remarkably similar structural pattern, a runaway exothermic reaction inside a sealed batch reactor, triggered by an interruption to normal cooling and agitation procedures, appears more than three decades earlier and an ocean away, at a chemical plant in Meda, near the town of Seveso, in northern Italy. On the morning of Friday, July 9, 1976, a batch of trichlorophenol, an intermediate chemical used in the production of disinfectant soap, began production roughly ten hours later than the plant's usual schedule inside a ten thousand liter reactor at the ICMESA chemical company. Early the following morning, Saturday, July 10, with the plant preparing to shut down for the weekend, operators stopped the reactor's heating supply while only about fifteen percent of the batch's solvent had been distilled off, a deviation from the standard completed process. Fifteen minutes later, they also switched off the reactor's stirring mechanism, and shortly after that, the technicians signed off and left the facility entirely unsupervised for the weekend, having brought the reactor down to atmospheric pressure but leaving its contents in a partially processed, chemically unstable state.

With no stirring to distribute heat evenly and no active cooling to remove the heat still being generated by the incompletely processed reaction mixture, the temperature inside the unsupervised, unattended reactor began to climb on its own. Scientific literature available even before the Seveso accident, and reportedly familiar to at least some of ICMESA's own technical staff, had already documented that this specific reaction mixture, a combination of caustic soda and ethylene glycol used in trichlorophenol production, could undergo an uncontrolled exothermic decomposition at elevated temperature, releasing large quantities of gas and, critically, forming dioxin as an unwanted and highly toxic byproduct of that decomposition pathway, a hazard the company's technical management had reportedly reviewed prior to the accident without translating that awareness into a change in the plant's actual unsupervised weekend shutdown procedure. At approximately 12:37 in the afternoon on Saturday, with the plant empty of its normal operating staff, the reactor's internal temperature and pressure exceeded the threshold at which a safety bursting disc was designed to rupture. Maintenance staff who happened to be nearby heard a whistling sound and watched a dense white cloud rise from a vent on the building's roof, an event that, notably, produced no fire and no explosion in the conventional sense, nothing resembling the violent detonation at T2 Laboratories or West, Texas, which is precisely why its true severity was not immediately recognized by anyone in the vicinity that afternoon.

The cloud that drifted from that vent over the following twenty minutes carried an estimated quantity of 2,3,7,8 tetrachlorodibenzo para dioxin, commonly known as TCDD, one of the most toxic synthetic compounds known to science, with subsequent scientific estimates of the total amount released ranging from a few hundred grams up to several kilograms, later research generally converging somewhere in the range of fifteen to thirty kilograms. Because the release produced no immediate visible catastrophe, no deaths occurred in the disaster's first hours or days, in stark contrast to T2, West, Texas, or the earlier chapters of this book. The consequences instead unfolded gradually and, in some respects, more insidiously: residents of Seveso and the surrounding towns began noticing unusual skin conditions, particularly a severe and disfiguring acne like condition called chloracne, especially among children who had been outdoors during and after the release, while local animals began dying in unusually large numbers within days, a warning sign that, tragically, took considerably longer for human authorities to fully interpret and act upon, in part due to significant early confusion and poor communication between the company and local government about the true scale and toxicity of what had been released. Twenty six pregnant women who had been exposed underwent abortions during the crisis, and the most heavily contaminated zones around the plant were eventually evacuated entirely, with the contaminated soil requiring years of remediation work.

Investigators who later examined the ICMESA accident in detail identified specific, correctable technical failures strikingly similar in character, if not in particular, to those later found at T2 Laboratories: reactor control systems that lacked adequate instrumentation to monitor the process's critical parameters in real time, no automated control system capable of intervening if those parameters drifted outside a safe range, and a pressure relief bursting disc set to activate only at a relatively high threshold of 3.5 bar, a design choice investigators later concluded meant that when the disc finally did rupture, it did so only after the reactor's contents had already reached a temperature high enough to generate substantial quantities of the unwanted and highly toxic dioxin byproduct, whereas a disc calibrated to rupture at a lower, less hazardous pressure and temperature might have vented the reactor's contents earlier in the runaway sequence, before as much dioxin had formed. In other words, even the emergency safety mechanism that ultimately prevented a Seveso style disaster from escalating into an outright explosion had itself been engineered without full consideration of the specific reactive hazard it was meant to guard against, a design gap conceptually identical to the educational and procedural gap the CSB later identified at T2, three decades apart, on two different continents, in two entirely different chemical processes.

The regulatory legacy of the Seveso disaster proved considerably more durable and far reaching than that of any single American grain elevator or fertilizer facility explosion discussed elsewhere in this book, in large part because the scale of environmental contamination and the involvement of an internationally recognized toxin gave the incident a political and scientific profile that a domestic industrial accident, however deadly, often struggles to achieve. The European Economic Community responded to Seveso by developing what became known as the Seveso Directive, first adopted in 1982, a binding legal framework requiring member states to identify facilities handling hazardous quantities of dangerous substances, mandate that those facilities conduct formal risk assessments, and establish safety management systems specifically designed to prevent major accidents and to limit their consequences to human health and the environment should prevention fail. The directive has been revised and strengthened multiple times in the decades since, and it remains, to this day, the foundational piece of chemical process safety regulation across the European Union, a direct institutional legacy of the specific failures, inadequate instrumentation, absent automated controls, and a pressure relief system not properly matched to the reaction it was meant to protect against, that investigators traced back to a single unattended reactor outside Milan on a July weekend in 1976.

Reading the T2 and Seveso cases together, separated by three decades, an ocean, ownership structures, chemical products, and regulatory environments, reveals a pattern that recurs throughout the industrial chapters of this book and that deserves to be stated as plainly as possible: a runaway exothermic reaction inside a sealed or partially sealed reactor is not primarily a chemistry problem in the sense of representing some poorly understood or unpredictable natural phenomenon. The chemistry, in both cases, was either already documented in the scientific literature before the accident occurred, as at Seveso, or well within the reach of standard chemical engineering education, as the CSB argued regarding T2. What failed, in both instances, was the translation of that available chemical knowledge into the specific engineering safeguards, adequate cooling redundancy, properly calibrated relief systems, real time process monitoring, and organizational procedures, adequate weekend supervision, a full appreciation among decision makers of what could go wrong, capable of reliably preventing a known hazard from becoming a realized one. This is, in the end, the most sobering lesson this chapter has to offer: neither Jacksonville nor Meda suffered from a mystery. Both suffered from a gap, well documented after the fact in both cases, between what the relevant chemistry could have told the people in charge and what those people, for reasons rooted in education, procedure, and the ordinary human tendency to underestimate a hazard that has not yet materialized, actually understood at the moment a decision needed to be made.

The specific engineering concept that connects these two disasters most directly, and that has become one of the central organizing principles of modern chemical process safety in the decades since, is the idea of a maximum credible event, a formal analytical exercise in which engineers deliberately imagine and quantify the worst plausible outcome of a given process failure before that process is ever put into operation, rather than discovering the answer only after an actual failure has occurred. Applied retroactively to T2 Laboratories, a maximum credible event analysis of the metalation reaction, asking specifically what would happen if cooling were lost entirely during the most thermally sensitive phase of the reaction, would very likely have identified the possibility of a runaway reaction capable of generating pressures the reactor vessel could not contain, precisely the outcome the CSB's investigation reconstructed after the fact using the same basic kinetic and thermodynamic principles that a rigorous pre operational hazard analysis would have applied in advance. Applied retroactively to the ICMESA reactor at Seveso, the same kind of formal analysis, asking what would happen if the reaction mixture were left unattended, unstirred, and without active cooling for an extended weekend period, would very likely have flagged the specific vulnerability that in fact materialized. Neither company appears to have conducted this kind of formal, structured worst case analysis with sufficient rigor before its respective disaster occurred, and the widespread adoption of formal reactive hazard review as a standard, often legally mandated component of chemical process design in the decades since both incidents represents one of the more concrete and lasting improvements to industrial safety practice that this book's case studies collectively illustrate.

It is worth pausing on the specific chemistry of the T2 metalation step, since it reveals a connection to an earlier chapter of this book that the general narrative of the disaster tends to obscure. The MCMT manufacturing process began by combining a chemical precursor called methylcyclopentadiene dimer with a solvent, and then introducing metallic sodium directly into the sealed reactor, the same alkali metal examined at length in this book's third chapter, chosen for this particular industrial process precisely because of its strong tendency to react with and displace other chemical groups, the identical underlying reactivity, a loosely held outer electron eager to transfer to a suitable partner, that makes sodium react so dramatically with water. Reactor procedures called for the sodium to be melted within the sealed vessel and combined with the methylcyclopentadiene under carefully controlled heating, with the process specifically designed so that heating would be discontinued once the mixture reached approximately three hundred degrees Fahrenheit, and active cooling engaged once the temperature reached three hundred sixty degrees, a narrow, deliberately engineered margin between the temperature at which the desired reaction proceeded efficiently and the temperature at which the CSB's investigation concluded the same reaction began accelerating into an uncontrolled and ultimately catastrophic thermal runaway. The reactor that exploded on December 19, 2007, had, according to the CSB's reconstruction, lost the cooling capacity specifically responsible for keeping the reaction within that narrow safe margin, allowing the sodium driven reaction to cross the threshold its own designers had identified, on paper, as the boundary between routine operation and genuine danger. This detail matters for this book's broader argument for a specific reason: it demonstrates that T2's own engineers evidently understood, at least in general terms, that this particular reaction carried a thermal runaway hazard serious enough to warrant a specifically defined temperature ceiling and an active cooling response protocol, and yet the CSB's investigation concluded that neither the reactor's engineering safeguards nor its emergency relief system had been sized with sufficient rigor to survive an actual, full loss of that cooling capacity during the reaction's most thermally sensitive phase, a gap between partial recognition of a hazard and full, quantified engineering protection against its worst case consequences that recurs, in one form or another, across nearly every industrial disaster examined in this book.

There is also a structural similarity in how both companies responded, or failed to respond, to warning signs that, in retrospect, appear to have been present well before the moment of catastrophic failure, and this pattern deserves attention because it recurs, in various forms, across nearly every disaster examined in this book. At Seveso, published scientific literature describing the specific decomposition and dioxin formation hazard associated with the plant's trichlorophenol process had been available to at least some of ICMESA's own technical staff prior to the accident, according to subsequent investigation, yet that documented awareness had not translated into a change to the specific unsupervised weekend shutdown procedure that ultimately allowed the runaway reaction to develop unmonitored. At T2 Laboratories, the CSB's investigation found no indication that the company had performed the kind of rigorous, documented reactive hazard evaluation that would have been expected given the exothermic nature of the metalation step, despite both owners holding relevant technical degrees and despite the broader chemical engineering literature containing ample documented precedent for exactly this category of hazard. In both cases, in other words, the gap was not between no information and some information, but between some available information and its full, rigorous, and organizationally embedded application to the specific process actually being run, a distinction this book has returned to so often across its industrial chapters that it deserves to be named plainly here: knowing that a hazard exists in the abstract and having fully internalized, operationalized, and defended against that specific hazard in a specific facility are two meaningfully different achievements, and the distance between them, again and again across the history this book has recounted, is where disaster has tended to live.

Chapter 9

The Battery in Your Pocket: Lithium and the Chemistry of Convenience

Most of the exothermic hazards examined so far in this book share a common feature that, on reflection, is quietly reassuring: they occur in settings most readers will never personally enter. Few people will ever stand near a fertilizer storage silo, operate an industrial batch reactor, or handle bulk quantities of reactive alkali metal. The subject of this chapter offers no such distance. It is very likely sitting within arm's reach of you right now, inside a phone, a laptop, a set of wireless earbuds, or, increasingly, an electric bicycle or scooter parked in a hallway or apartment stairwell. The lithium ion battery has become one of the most quietly ubiquitous exothermic hazards in modern life, not because it is poorly designed or unusually dangerous under normal operation, but because its failure mode, when it does fail, is a particularly aggressive and fast moving example of the thermal runaway dynamic this book has returned to again and again.

A lithium ion battery cell is, in essence, a carefully engineered chemical system held in a deliberately unstable, energy dense configuration. Inside each cell sit two electrodes, a cathode and an anode, separated by a thin physical barrier and immersed in a liquid electrolyte that allows lithium ions to move back and forth between the electrodes during charging and discharging, which is how the battery stores and releases electrical energy. This design is, by its very nature, a compromise between energy density and stability, since packing more usable energy into a smaller, lighter cell, the design goal that has driven decades of battery improvement and made modern portable electronics and electric vehicles practical in the first place, generally means thinner separators, more reactive electrode chemistries, and less margin for error if something inside the cell goes wrong. Under normal use, this compromise works extremely well, and the vast majority of lithium ion batteries manufactured and used worldwide complete their functional lives without ever experiencing a safety incident. The concern this chapter addresses applies specifically to the minority of cases in which something disrupts that careful internal balance.

Battery safety researchers generally categorize the triggers for lithium ion thermal runaway into three broad types: mechanical abuse, such as a puncture, crush, or sharp impact that physically damages the cell's internal structure; electrical abuse, such as overcharging, over discharging, or exposure to a short circuit; and thermal abuse, meaning exposure to external heat sufficient to destabilize the cell's internal chemistry directly. Each of these triggers accomplishes the same underlying result by a different path: damage to or breakdown of the thin separator layer that normally keeps the cell's cathode and anode physically apart. Once that separator fails, even in a small localized area, the cathode and anode can come into direct or near direct contact, creating an internal short circuit. Unlike an external short circuit, which a circuit breaker or fuse can often interrupt, an internal short circuit occurs entirely inside the sealed cell, generating an intense, immediate, and completely uncontrolled surge of heat at exactly the location where the separator failed, heat that has nowhere to safely dissipate because it is generated inside the very structure meant to contain the battery's stored energy.

This is where the underlying chemistry of the battery itself takes over and begins actively working against containment rather than merely failing passively. The heat generated by the initial internal short circuit is sufficient, once it reaches a critical threshold, to begin breaking down a protective chemical layer on the surface of the anode known as the solid electrolyte interface, and this breakdown is itself an exothermic reaction, one that releases additional heat directly into the same small, enclosed space where the original short circuit occurred. That additional heat accelerates the breakdown of surrounding materials further, which releases still more heat, in a self reinforcing cascade that follows precisely the exponential Arrhenius relationship between temperature and reaction rate introduced earlier in this book, compressed here into a timeframe of seconds rather than the hours or days seen in a hay stack or a compost pile. As internal temperature climbs into the hundreds of degrees Celsius, the cell's liquid electrolyte itself begins to decompose, generating flammable gases inside the sealed cell casing. Pressure builds rapidly alongside temperature, and the combination frequently results in the cell venting forcefully, sometimes rupturing outright, releasing hot gas, sparks, and burning electrolyte, and very often igniting on contact with the surrounding air, a sequence that researchers studying the phenomenon describe, accurately, as an exothermic chain reaction capable of pushing internal cell temperature up by thousands of degrees within a very short span of time.

The scale of this hazard, once confined mostly to occasional, well publicized incidents involving laptop batteries or early generation electric vehicles, has grown considerably alongside the explosive popularity of lithium ion powered micromobility devices, particularly electric bicycles and electric scooters, many of which use less rigorously regulated battery packs, chargers, or aftermarket replacement batteries than the devices' original manufacturers intended. New York City, which has seen an especially rapid rise in e bike and e scooter adoption for food delivery and personal transportation, offers the most closely tracked and most instructive data on this hazard available anywhere, precisely because the city's fire department has invested heavily in monitoring the problem year over year. Fire department data showed 277 fires attributed to lithium ion batteries in 2024, a modest increase from 268 the year before, yet over that same period the number of deaths connected to these fires fell sharply, from eighteen fatalities in 2023 to six in 2024, a reduction of roughly two thirds achieved in a single year without any fundamental change to the underlying battery chemistry itself. Fire officials attributed this improvement primarily to a sustained public education campaign encouraging residents to charge and store these devices outdoors wherever possible, reflected in the data by a rise in the proportion of incidents classified as non structural, meaning the fire occurred outdoors rather than inside an apartment or business, from ninety such fires in 2023 to one hundred thirty three in 2024, evidence that a comparatively simple behavioral change, moving a charging battery a few feet outside a living space, meaningfully altered the consequences of a hazard that the underlying chemistry made effectively impossible to eliminate outright.

One detail from the city's public safety messaging deserves particular emphasis, since it corrects a common and potentially dangerous misconception about when these fires actually occur. Following a fire at an e bike retail store in Queens in August 2024, the fire department's public messaging specifically and deliberately emphasized that lithium ion batteries do not need to be actively charging in order to fail catastrophically, directly countering an intuitive but mistaken assumption that the danger window is limited to the hours a device sits plugged in. Any of the three general failure categories introduced earlier in this chapter, mechanical damage from a drop or collision, an internal manufacturing defect of the kind examined in the Note7 case, or gradual degradation from repeated charge cycles, can trigger thermal runaway in a battery that is simply sitting idle, fully charged and disconnected from any power source, a fact that meaningfully expands the scope of reasonable precaution beyond simply avoiding overnight charging and toward more comprehensive practices around device storage, handling, and the willingness to retire a battery once it shows any sign of physical damage regardless of whether it happens to be plugged in at that particular moment.

The problem is not confined to ground level transportation. UL Standards and Engagement, an organization that tracks lithium battery incidents specifically within commercial aviation through a dedicated monitoring program, documented an average of two thermal runaway incidents reported by airlines every week during 2024, a rate essentially unchanged from the prior year and among the highest recorded in six consecutive years of tracking, with the overwhelming majority of these incidents, roughly ninety percent, occurring with devices actually aboard the aircraft rather than in checked or cargo baggage, a distinction that matters considerably for how quickly a developing thermal runaway event can be identified and addressed by crew before it escalates. The specific devices involved in aviation incidents skew toward the same category of consumer electronics most people carry as a matter of routine, phones, laptops, tablets, and increasingly portable power banks, rather than any exotic or specialized equipment, underscoring that this is fundamentally a hazard embedded in extremely common, everyday objects rather than one confined to unusual or high risk equipment.

Understanding why lithium ion thermal runaway is difficult to stop once it begins, rather than simply difficult to prevent in the first place, is important for calibrating an appropriately serious response to the hazard without tipping into unhelpful panic about devices most people rely on daily and cannot reasonably avoid. Once a cell enters thermal runaway, conventional firefighting approaches that work well against ordinary combustible fires are frequently far less effective, because the battery's own internal chemical reaction, not simply the burning of external material, continues generating heat and can reignite even after visible flames have apparently been extinguished, particularly in multi cell battery packs where thermal runaway in one cell can propagate heat sufficient to trigger runaway in adjacent cells, a cascading failure mode that battery engineers refer to as thermal propagation and that represents one of the most active areas of contemporary battery safety research, since preventing propagation between cells, even when an individual cell failure cannot be entirely eliminated, dramatically reduces the severity of the resulting fire.

This engineering research effort is worth dwelling on briefly, because it represents one of the more genuinely encouraging threads running through this book's otherwise sober subject matter. Battery manufacturers and safety researchers have made substantial, measurable progress on several fronts simultaneously: improved separator materials designed to resist puncture and to fail in ways that minimize the area of direct electrode contact if they do fail; battery management systems capable of detecting the earliest thermal and electrical signatures of an developing internal fault and shutting a cell down before runaway fully develops; physical pack designs that incorporate thermal barriers specifically intended to prevent heat from one failing cell from triggering its neighbors; and, perhaps most fundamentally, alternative battery chemistries, particularly lithium iron phosphate formulations, which offer somewhat lower energy density than the cobalt based chemistries common in early consumer electronics but which are considerably more thermally stable and less prone to runaway under mechanical or electrical abuse, a tradeoff an increasing number of electric vehicle and stationary storage manufacturers have judged worthwhile given the safety benefit.

For the ordinary consumer, the practical takeaways from the underlying chemistry are fairly specific and considerably more actionable than the phrase battery safety might initially suggest. Devices and batteries should be charged using chargers designed and rated specifically for that device, since mismatched voltage or current delivery is a well documented pathway into the electrical abuse category of thermal runaway triggers. Visibly damaged batteries, swollen, punctured, or subjected to a hard impact or drop, should be treated as potentially compromised even if they still appear to function normally, since internal separator damage is frequently invisible from the outside until the moment it fails. Charging unattended overnight, particularly for higher capacity devices like e bikes and e scooters, concentrates risk precisely during the hours when a developing fire is least likely to be noticed quickly, which is why many fire safety authorities now specifically recommend charging such devices away from primary escape routes and, where possible, while someone remains awake and present nearby. None of this eliminates the hazard entirely, since even properly manufactured, properly charged, undamaged lithium ion cells retain some small residual failure rate inherent to the chemistry itself. What careful handling does accomplish, consistent with the broader argument of this book, is meaningfully lowering the odds that any particular cell ever receives the specific mechanical, electrical, or thermal nudge required to cross from stable, useful energy storage into the self accelerating exothermic cascade that battery engineers, echoing chemists a century before them, still call, for good reason, a runaway.

One of the most instructive illustrations of how a manufacturing defect, rather than any misuse by the end consumer, can push a mass produced battery across that threshold occurred in the autumn of 2016, when Samsung Electronics launched its flagship Galaxy Note7 smartphone to considerable commercial anticipation, only to recall the device entirely within weeks after reports began accumulating of units overheating, smoking, and in a number of documented cases catching fire. The United States Consumer Product Safety Commission, in its formal recall notice, documented ninety two reports of battery overheating within the United States alone, including twenty six reports of burns and fifty five reports of property damage, among them fires that damaged vehicles and a residential garage, figures that reflected only the initial wave of incidents before the scale of the problem became fully apparent. Samsung's subsequent internal investigation, which the company reported involved roughly seven hundred researchers and engineers testing more than two hundred thousand devices and more than thirty thousand individual batteries in an effort to replicate and fully understand the failure, ultimately identified two separate and entirely distinct manufacturing defects, originating with two different battery suppliers, each independently capable of triggering thermal runaway through the same fundamental separator failure mechanism described earlier in this chapter, arrived at by two different manufacturing pathways.

The first defect, found in batteries supplied by one manufacturer, traced back to a battery casing design that Samsung's investigation concluded was slightly too small for the internal battery components it was meant to house, a mismatch that caused the negative electrode inside the cell to bend under the resulting mechanical pressure, increasing the risk that the electrode would eventually breach the separator and trigger an internal short circuit, precisely the separator failure mechanism this chapter has described as the common pathway into thermal runaway regardless of its specific triggering cause. The second defect, found in batteries from an entirely different manufacturer brought in specifically to supply replacement units after the first recall, arose from a different but conceptually related manufacturing flaw: sharp burrs, microscopic protrusions of metal left behind during the welding process used to assemble the battery's internal components, which under the ordinary mechanical stress of ongoing expansion and contraction that lithium ion cells experience during normal charging and discharging cycles, gradually worked their way through the cell's internal insulation layers, again producing exactly the kind of internal short circuit capable of initiating thermal runaway. The bitter irony embedded in this second defect, that it affected the very batteries Samsung had shipped specifically as safe replacements for the first, already recalled defect, meant that a significant number of consumers who had dutifully exchanged their original Note7 for what they had been assured was a corrected, safe replacement device found themselves holding a phone with an entirely different but equally dangerous manufacturing flaw, a sequence of events that transformed what might otherwise have been remembered as a single, contained product safety incident into what industry analysts have since described as one of the most costly and reputationally damaging consumer electronics recalls in recent history, with Samsung's own estimate of the total financial impact reaching more than five billion dollars.

The Note7 episode is worth including in this book not for its financial or reputational consequences, considerable as those were, but because it demonstrates with unusual clarity a point this book has made in several of its industrial case studies as well: a hazard rooted in well understood, thoroughly documented chemistry can still slip past an enormous, well resourced, and genuinely sophisticated engineering and quality assurance organization, not because the underlying thermal runaway mechanism was mysterious or unknown to Samsung's battery engineers, but because a specific manufacturing tolerance, a casing dimension measured in fractions of a millimeter, or a microscopic welding burr invisible without specialized inspection equipment, proved sufficient to breach the razor thin margin of physical separation standing between a lithium ion cell's electrodes and a catastrophic internal short circuit. Independent researchers examining the broader phenomenon of lithium battery failure have noted that even the smallest imaginable spark inside a cell is sufficient to initiate the thermal runaway cascade, a statement that, read alongside the Note7 investigation's findings, underscores just how demanding the manufacturing precision required to reliably avoid that outcome across millions of mass produced units genuinely is, and why battery safety engineering, even in devices from the most experienced and best resourced manufacturers, remains an area of continuous refinement rather than a solved problem.

Chapter 10

Kitchen Chemistry: Deep Fryers, Grease Fires, and the Physics of Cooking Oil

Cooking is, by a wide margin, the leading cause of home fires in the United States, and it holds that grim distinction not because kitchens are inherently dangerous places but because ordinary cooking oil, left unattended on a hot burner, is fully capable of igniting entirely on its own, without a spark, a flame, or any external ignition source whatsoever, purely as a function of temperature. According to National Fire Protection Association data covering the years 2017 through 2021, cooking accounted for 44 percent of all reported home fires, 18 percent of home fire deaths, and 40 percent of home fire injuries during that period, and the great majority of these incidents trace back to a single, specific, and thoroughly avoidable cause: unattended frying, oil left heating on a stovetop while the cook stepped away, even briefly, for what seemed at the time like an entirely reasonable interval.

The chemistry behind this hazard is, at its core, a story about oxidation, the same general chemical process discussed in earlier chapters regarding hay and coal, but occurring here across a matter of minutes rather than weeks. Cooking oils and fats are composed primarily of fatty acid esters, and when heated in the presence of atmospheric oxygen, these molecules undergo oxidation reactions that release heat as a byproduct. At moderate cooking temperatures, this oxidation proceeds slowly enough that the released heat simply dissipates into the surrounding air, the pan, and the food being cooked, without any risk of runaway. But as an unattended pan of oil continues heating, unbuffered by food or active stirring, the oil first reaches what is known as its smoke point, generally somewhere between roughly 300 and 480 degrees Fahrenheit depending on the specific type of oil, refined oils with higher smoke points than unrefined or less processed varieties, at which point the oil begins visibly smoking as some of its components start to break down. Continue heating past the smoke point, and the oil reaches its flash point, the temperature at which vapors rising from its surface can be ignited by an external flame if one happens to be present, typically a stovetop's open gas burner sitting immediately below the pan. And if the oil is allowed to continue heating even further, unattended and with no ignition source required at all, it will eventually reach its autoignition temperature, generally falling somewhere between 700 and 815 degrees Fahrenheit for most common cooking oils, the exact point at which the oxidation reaction taking place within the oil itself generates enough heat, fast enough, to sustain open combustion without any spark, flame, or external trigger whatsoever. Research conducted by fire protection engineers at the University of Maryland, examining the specific autoignition behavior of a range of common cooking oils, butter, and margarine, found that depending on the specific fat involved and the heat source, it can take as little as eighteen minutes of unattended heating for a pan of oil to autoignite entirely on its own, a strikingly short window that helps explain why so many home cooking fires begin during precisely the kind of brief absence, answering a phone call, checking on a child, stepping into another room, that feels, in the moment, like an entirely reasonable and low risk decision.

Once a grease fire has started, the instinctive human response, reaching for water, the substance most people associate with extinguishing fire in every other context, is catastrophically wrong, and understanding exactly why requires returning to basic physics rather than chemistry. Burning cooking oil sits at a temperature well above five hundred degrees Fahrenheit, while water, of course, boils at a mere 212 degrees Fahrenheit at standard atmospheric pressure. When even a small quantity of water makes contact with oil at that temperature, the water does not simply extinguish the flame the way it might smother a fire involving ordinary combustible solids like wood or paper. Instead, the water flash boils almost instantaneously, converting from liquid to vapor and expanding, in the process, to roughly seventeen hundred times its original volume within a fraction of a second. This explosive expansion does not happen gently. It violently ejects burning oil in every direction, launching droplets of flaming grease upward and outward with enough force to reach a kitchen ceiling and to spread what had been a contained pan fire across an entire kitchen, and potentially onto the person standing nearby, within seconds. Fire safety data bears this mechanism out in grim detail: more than half of the civilians injured in cooking related home fires are hurt specifically while attempting to extinguish the fire themselves, a statistic that reflects, in significant part, exactly this kind of well intentioned but chemically catastrophic response.

The same reasoning explains why several other seemingly sensible instincts are equally dangerous when applied to a grease fire specifically, even though they might work reasonably well against other categories of household fire. Carrying a flaming pan to a sink, an instinct some home cooks report having seen modeled, unfortunately, even on television cooking programs, risks sloshing burning oil onto the mover's own body during transport and very often ends with the pan reaching a sink equipped with a water tap, recreating the same steam explosion hazard in an even more poorly controlled setting. Standard multipurpose dry chemical fire extinguishers, the kind most commonly kept in American kitchens and rated for ordinary combustible, electrical, and flammable liquid fires generally, discharge their contents at relatively high pressure, a design feature that works well against many fire types but that, aimed directly at a shallow pan of burning oil, can blast the burning liquid up and out of the pan entirely, scattering it across the kitchen in a manner not unlike the water and steam mechanism, just without the added violence of a true steam explosion. Carbon dioxide extinguishers carry a similar risk of disturbing and scattering the burning oil through the sheer force of their discharge.

The specific tool fire safety authorities now recommend for cooking oil and grease fires, a designation that did not always exist as a distinct category, is the Class K extinguisher, so named for the specific hazard class covering cooking media fires that the National Fire Protection Association formally recognized as requiring separate treatment from ordinary flammable liquid fires. Class K extinguishers work through an entirely different chemical mechanism than either water or standard dry chemical extinguishers. They discharge a fine mist of a wet chemical agent, typically potassium acetate based, applied at deliberately low pressure specifically to avoid disturbing or splashing the burning oil's surface. Upon contact, this agent reacts with the hot oil to form a thick, soapy foam layer across the entire surface, a process called saponification, essentially the same basic chemistry behind traditional soap making, using heat and a base to convert fats into soap. This foam layer accomplishes two things simultaneously that neither water nor a standard extinguisher can manage together: it smothers the fire by excluding oxygen from the oil's surface, and its water content gradually absorbs heat from the oil beneath it, cooling the entire mass over time down below its autoignition temperature, which matters enormously because burning cooking oil has a well documented tendency to reignite even after an initial fire appears extinguished, since the oil beneath a doused surface often remains well above its autoignition point and will resume burning as soon as fresh oxygen reaches it again, a phenomenon fire investigators refer to as reflash and a major reason why professional grade Class K systems, required by fire code in commercial kitchens under National Fire Protection Association standards, are specifically engineered to maintain suppression over an extended period rather than achieving only a quick, superficial knockdown of visible flame.

For a home cook facing a grease fire without a Class K extinguisher on hand, which describes the overwhelming majority of American households, fire safety authorities recommend two specific, low technology responses that work by the same basic principle of oxygen exclusion without the added risk of disturbing the burning oil. Sliding a metal lid over the pan, carefully and from the side rather than directly from above, cuts off the fire's oxygen supply and, left in place with the burner turned off, will generally smother a contained grease fire within a short period, though the lid should be left undisturbed for a considerable time afterward given the reflash risk described above. A purpose made fire blanket, a household safety item that remains far less commonly owned in the United States than it is in many European countries but that fire safety educators increasingly recommend keeping within easy reach of any stovetop, accomplishes the same oxygen exclusion goal for pans too large or awkwardly positioned for a simple lid, and can also be used to safely smother flames on a person's clothing in the rare event that a kitchen fire spreads that far.

There is a broader thematic point embedded in this chapter that connects it back to the wider argument of this book despite its comparatively modest, domestic scale relative to a fertilizer warehouse explosion or an industrial reactor rupture. The grease fire is, in miniature, a nearly perfect illustration of the gap between chemical intuition and chemical reality that runs through every chapter of this book. Water is, in the vast majority of everyday contexts, the correct and instinctive response to fire, and that instinct has been reinforced by a lifetime of ordinary experience with campfires, candles, and burning paper, none of which behave anything like a pan of oil heated past its autoignition point. The specific chemistry of cooking oil, its capacity to ignite entirely on its own without any external spark, and the violent, physics driven consequences of introducing water to a liquid burning at more than double water's boiling point, sits entirely outside that everyday intuition, which is exactly why so many people, including some who cook daily and consider themselves reasonably careful in the kitchen, continue to reach for water, or for a standard extinguisher, or for the sink, in precisely the moment when doing so causes far more harm than the fire would have caused on its own if simply smothered and left alone.

A related but distinct phenomenon, one that connects a kitchen grease fire to some of the deadliest building fires in modern history, is worth understanding here as well, because it explains why a fire that begins as something small and apparently contained can transform, within a matter of seconds, into something that fills an entire room simultaneously. Fire scientists call this transition flashover, the point at which the accumulated heat within an enclosed space becomes sufficient to raise every combustible surface in that space, walls, furniture, carpeting, ceiling material, to its own ignition temperature at essentially the same moment, so that a fire that had been confined to one corner of a room abruptly becomes a fire consuming the entire room at once. The physical mechanism responsible involves exactly the exothermic feedback dynamic introduced in this book's opening chapter: as a fire burns, it heats the air and surfaces around it, and that heat radiates outward and accumulates within an enclosed space rather than dispersing freely the way it would outdoors, gradually raising the temperature of every combustible material in the room even before those materials show any visible sign of burning. Once that accumulated heat crosses the ignition threshold for the materials present, everything ignites nearly simultaneously, and the resulting flashover can transform a room from a fire in one corner to a room fully engulfed in flame in a matter of seconds, far too quickly for anyone still inside to reliably escape.

The MGM Grand Hotel fire in Las Vegas, on the morning of November 21, 1980, remains one of the starkest illustrations of flashover's speed and lethality on record. The fire began with a comparatively minor electrical ground fault inside a concealed space at a restaurant serving station, the kind of small, contained ignition source that, in a room furnished with genuinely noncombustible materials, might have smoldered briefly and self extinguished without ever being noticed. Investigators later found that the casino's furnishings, however, included extensive plastic and foam materials, chairs and booths padded with combustible foam, decorative elements built from rigid plastics designed to visually resemble wood, that provided exactly the kind of fuel load capable of sustaining and accelerating a developing fire. Once the initial fire reached flashover conditions within the restaurant area, investigators documented that the flame front swept across the adjoining casino floor at a rate of roughly six feet per second, with virtually every combustible surface in its path igniting within twenty seconds of the flame front's arrival, a speed of propagation that left many of the casino's occupants, positioned at gaming tables and slot machines, with almost no warning and almost no time to reach an exit before the space around them became entirely engulfed. Eighty five people died and nearly seven hundred were injured, the great majority from smoke inhalation rather than direct burns, since the fire's smoke and toxic combustion products spread through the hotel's ventilation system considerably faster and considerably further than the flames themselves.

The disaster prompted sweeping changes to fire and building codes across the United States, including significantly stricter requirements governing the flammability of interior furnishings in public buildings and mandatory sprinkler installation in high occupancy structures, changes explicitly designed to address the specific vulnerability the MGM Grand fire exposed: that a modern building's interior, filled with synthetic materials selected for appearance and cost rather than for fire resistance, could support a flashover event far more readily than the same space furnished with more traditional, less combustible materials might have. A strikingly similar dynamic, involving flashover driven by combustible interior finishing rather than furniture, played out more than three decades later and on another continent, at the Kiss nightclub in Santa Maria, Brazil, where a pyrotechnic device used during a live musical performance ignited acoustic foam installed across the venue's ceiling on January 27, 2013. The foam, selected for its sound dampening properties rather than any consideration of fire behavior, ignited and spread flame across the ceiling with extraordinary speed, and the resulting fire, compounded by a lack of clearly marked emergency exits and a power failure that left much of the venue in darkness, killed two hundred forty two people, the majority again through smoke inhalation in a crowd unable to locate or reach a limited number of exits quickly enough.

Both disasters belong in this book for a reason that extends beyond their shared involvement of exothermic combustion chemistry: each demonstrates, with tragic clarity, how the ordinary materials that fill built environments, furniture foam, decorative plastic, acoustic insulation, chosen for entirely reasonable aesthetic, comfort, or functional reasons, can transform a small, containable ignition event into a flashover capable of engulfing an entire room within seconds, precisely the kind of exothermic acceleration this book has traced across settings as varied as a hay barn, a battery cell, and an industrial reactor, here playing out within the walls of spaces millions of people pass through as a matter of ordinary daily life, largely unaware that the specific materials surrounding them have any bearing at all on how a fire, should one ever start, would actually behave.

The history of the Class K extinguisher category itself offers a useful window into how slowly institutional fire safety standards can sometimes catch up to well established kitchen chemistry, a pattern that will be familiar by now to readers who have made their way through this book's earlier chapters on grain dust and ammonium nitrate regulation. Commercial kitchen fire suppression prior to the widespread adoption of dedicated wet chemical systems relied primarily on dry chemical extinguishers originally developed and standardized for a considerably broader range of flammable liquid fires, extinguishers that, while reasonably effective against many fuel types, were never specifically engineered around the particular combination of very high sustained cooking oil temperatures and the strong tendency toward reflash that this chapter has already described. The National Fire Protection Association's development of a distinct Class K designation, and the corresponding engineering standards for wet chemical suppression systems specifically designed around the saponification mechanism, emerged only as accumulated fire data from commercial kitchens made clear that a fire category common enough to represent the single leading cause of restaurant fires warranted its own dedicated, chemistry specific suppression technology rather than continued reliance on general purpose equipment adapted, imperfectly, to a highly specific hazard.

It is worth pausing on the deep fryer specifically, since commercial and industrial deep frying operations represent something of an intensified, higher stakes version of the same hazard examined in this chapter's opening sections, and illustrate how scale changes consequence even when the underlying chemistry remains unchanged. A large capacity commercial fryer may hold tens of gallons of oil at any given time, heated continuously throughout a restaurant's operating hours and subjected to repeated heating and cooling cycles across each day's service, cycles that fire safety engineers note can accelerate the oil's own chemical degradation over time, gradually lowering its effective flash point as breakdown byproducts accumulate within the oil itself, a phenomenon distinct from but conceptually related to the aging chemistry discussed elsewhere in this book regarding organic peroxides slowly becoming more hazardous the longer they sit in imperfect storage conditions. This is part of why commercial kitchen fire codes generally mandate not only Class K suppression systems but also automatic shutoff mechanisms tied directly to the fryer's heating element, along with regular oil testing and replacement schedules, a layered set of engineering and procedural controls that mirrors, in a much smaller and more domestic setting, the same defense in depth philosophy examined at industrial scale in this book's chapter on reactive hazard prevention.

For the home cook, the takeaway from all of this technical detail is refreshingly simple, and worth stating plainly as this chapter draws to a close: cooking oil is not dangerous in the way that a stored industrial chemical or a reactive metal is dangerous, requiring specialized handling knowledge for ordinary use. It becomes dangerous specifically and only when heated unattended for long enough to approach its autoignition threshold, and the single most effective safety practice available to any home cook, more effective by a wide margin than owning any particular extinguisher or safety device, is simply never leaving oil unattended on an active heat source, a habit that costs nothing, requires no specialized equipment, and directly addresses the one condition, extended unattended heating, that every documented cooking fire statistic in this chapter traces back to as the common root cause.

Chapter 11

The Peroxide Problem: Organic Peroxides and the Hazards of Aging Chemicals

In 1995, at the University of California, San Francisco, a laboratory worker conducting a routine cleanout placed two cans of diethyl ether, untouched since 1985, into a fume hood and began venting their lids, a standard precaution before disposal. He heard a single, sharp pop and turned toward the sound just in time to be knocked down by the resulting blast. A nearby bottle of nitric acid, disturbed by the same explosion, shattered as well, spraying acid onto his face and hair. Nobody had opened those cans in a decade. Nobody had done anything to them at all in ten years, which is precisely the point of this chapter: sometimes the most dangerous thing a chemical container can do is simply sit, sealed and undisturbed, on a shelf, slowly becoming something considerably more hazardous than whatever was originally poured into it.

The mechanism responsible is a slow chemical process called autoxidation, and diethyl ether, along with a specific family of related solvents including tetrahydrofuran, isopropanol, dioxane, and isoamyl alcohol, is particularly prone to it. These compounds share a chemical structure that reacts, gradually and without any external trigger beyond ordinary exposure to atmospheric oxygen and light, to form organic peroxides within the liquid itself. This reaction proceeds slowly enough that a freshly opened, properly stored bottle of ether poses essentially no elevated risk, and safety guidance from institutions handling these solvents regularly generally holds that a full, unopened, amber glass bottle stored cool and out of direct light can sit safely for as long as a year without accumulating dangerous peroxide levels. The danger accumulates precisely in the scenario that seems, intuitively, like it should be the safest: a bottle that has been opened once, perhaps used for a small quantity of solvent, resealed, and then left largely undisturbed for months or years, during which time continued exposure to trace atmospheric oxygen inside the container allows peroxide concentration to build slowly and steadily, entirely invisible from the outside, since the ether itself typically shows no obvious change in appearance, smell, or behavior even as its peroxide content climbs toward dangerous levels.

What makes accumulated peroxides so hazardous is not merely their presence but their behavior under specific, common laboratory conditions, and understanding this requires connecting back to the concept of activation energy introduced in the opening chapter of this book. Organic peroxides are, chemically speaking, compounds that are considerably more willing to release their stored energy suddenly and violently than the parent solvent they formed within, and critically, the activation energy required to trigger that release is often extremely low, low enough that mechanical shock, friction, or even a sudden temperature change can provide sufficient nudge to trigger detonation, without any spark or flame required at all. Safety guidance issued after documented university laboratory incidents describes this danger with a specific, almost unsettling precision: shocks as small as unscrewing a cap containing peroxide residue have proven sufficient to trigger detonation in properly documented cases, which is why laboratory safety officers generally treat any unlabeled, undated, or long stored container of a known peroxide forming solvent with the same level of caution normally reserved for a known explosive, rather than as an ordinary bottle of laboratory solvent that merely happens to be somewhat old.

The specific danger multiplies considerably during a process called concentration, and this is the mechanism behind several of the most serious documented peroxide accidents in academic and industrial laboratories. Peroxides, unlike the solvent they form within, generally do not evaporate as readily, which means that if a peroxide contaminated solvent is distilled, or allowed to slowly evaporate, or processed through a piece of laboratory equipment called a rotary evaporator that is specifically designed to remove solvent from a mixture under reduced pressure, the ether or tetrahydrofuran evaporates away while the peroxide compounds remain behind, becoming progressively more concentrated as the surrounding liquid volume decreases. A well documented incident at the University of California, Berkeley, illustrates exactly this mechanism in stark detail. A researcher was using tetrahydrofuran, a common laboratory solvent, in a reaction, and this particular batch of tetrahydrofuran had not been formulated with a stabilizing additive, typically a compound called BHT, that is specifically included in many commercial solvent formulations to slow the rate of peroxide formation during storage. Testing conducted after the fact found the solvent contained peroxide levels exceeding one hundred milligrams per liter, a substantially elevated concentration. As the rotary evaporator did its job, steadily removing the tetrahydrofuran itself, the peroxide compounds concentrated in the bottom of the glass reaction vessel, eventually drying into shock sensitive crystals. Any subsequent movement of that vessel was, at that point, sufficient to trigger detonation, and detonation is exactly what occurred, sending glass fragments into the researcher's face, protected in that instance, though not entirely, by safety goggles.

A comparable incident, differing mainly in its specific circumstances rather than its underlying chemistry, occurred at the University of Edinburgh, where an unopened, nearly full two liter bottle of diethyl ether exploded while sitting inside a solvent storage cupboard, having been stored under cool, dark conditions, generally considered the safest available storage practice, for only about seven months before detonating with no one present to trigger it through any obvious physical disturbance. Investigators concluded that peroxide formation was the most probable cause, a finding notable specifically because it undercuts the comforting assumption that unopened containers, stored under textbook correct conditions, are inherently safe from this hazard. As one university safety guidance document on the subject notes with particular emphasis, an ether bottle that appears empty is very often not actually empty in any meaningful sense, since it remains filled with ether vapor unless that vapor has been deliberately displaced, and a bottle containing a mixture of ether vapor and air represents close to ideal conditions for continued peroxide formation, a bottle that a casual observer might reasonably assume poses minimal risk precisely because it looks, from the outside, like it contains almost nothing at all.

The broader category of peroxide forming chemicals extends considerably beyond laboratory ether and tetrahydrofuran, and includes compounds like benzoyl peroxide, widely used industrially as a polymerization initiator and, in a considerably more diluted and stabilized form, as the active ingredient in many over the counter acne treatments, a striking illustration of how the same underlying chemical reactivity that makes a compound industrially useful, in benzoyl peroxide's case, its willingness to decompose and generate free radicals that can initiate other chemical reactions, can also make concentrated or improperly stored quantities of the same basic compound genuinely hazardous. Laboratory and industrial safety protocols accordingly classify peroxide forming chemicals into distinct hazard categories based on how readily and how dangerously they form these compounds, with some, classified at the highest hazard tier, capable of forming explosive peroxide concentrations even while still sealed in their original, unopened containers, simply as a function of elapsed time, while others form peroxides more slowly or only after being opened and exposed to air, and require testing and monitoring on a more moderate but still regular schedule.

This is where the chapter's central theme connects most directly to the broader argument running through this entire book. Every hazard examined in earlier chapters, from a hay stack to a battery to a fertilizer silo, involves an exothermic process that is, in some sense, actively developing, generating heat, building pressure, accumulating risk through an identifiable and generally monitorable physical or chemical process over a defined and often bounded period. The peroxide hazard is subtly different, and in some ways more insidious, precisely because the container itself, sitting sealed and undisturbed on a shelf, gives absolutely no external indication that anything is happening inside it at all. There is no rising smell, no visible discoloration in most cases, no gradually warming surface a careful worker might notice and investigate. The bottle looks, on the day it is finally rediscovered during a laboratory cleanout, exactly as unremarkable as it looked the day it was first opened, possibly a decade or more earlier, and it is precisely that appearance of total, unchanging inertness that makes it so dangerous, because nothing about the container's external presentation offers any warning that its activation energy for violent decomposition has, over the intervening years, quietly and invisibly fallen to a level low enough that the simple act of twisting off its cap could be sufficient to trigger an explosion.

The institutional response to this specific and well documented hazard has, appropriately, centered less on any new chemistry, since the autoxidation reactions responsible have been understood for many decades, and more on the unglamorous discipline of inventory management. Laboratory safety programs at universities and industrial facilities that handle known peroxide forming chemicals now generally require that every container be labeled, at the moment it is first received and again at the moment it is first opened, with the specific date, since the safe storage window for an opened container of a peroxide forming solvent is measured in months rather than years for many common compounds. Regular testing protocols, using simple chemical test strips capable of detecting elevated peroxide concentrations well before they reach genuinely dangerous levels, are standard practice at institutions that have taken the hazard seriously, and containers that have exceeded their documented safe storage window, or that show visible crystal formation, an unusually viscous texture, or any other sign of possible peroxide accumulation, are treated by trained hazardous materials personnel using specialized handling procedures rather than disposed of through ordinary laboratory waste channels. None of these measures require any advance in chemical understanding. They require only the same discipline this book has identified as the limiting factor again and again across a hay barn, a grain elevator, and a fertilizer warehouse: the sustained, unglamorous, easily deferred practice of checking on something that, by every outward appearance, looks like it does not need checking on at all.

The broader chemical family of peroxide forming compounds extends well beyond laboratory solvents and includes another category of substance familiar to a considerably wider public than diethyl ether or tetrahydrofuran: certain plastics and monomers used in everyday consumer and industrial products. Styrene, the chemical precursor to polystyrene, one of the most widely produced plastics in the world, is itself a peroxide former under sufficiently prolonged storage, particularly at elevated temperature or in the presence of light, which is why commercial styrene shipments are almost always formulated with an inhibitor additive specifically intended to slow this process during the transport and storage period between manufacture and use. Acrylic acid, methyl methacrylate, and a range of other industrially significant monomers share this same underlying vulnerability, and industrial chemical storage guidance for these substances consistently emphasizes the same core principles established in laboratory settings, dated inventory, temperature controlled storage, regular testing, and firm maximum storage durations, scaled up from a laboratory shelf holding a handful of bottles to an industrial tank farm holding thousands of gallons, where the consequences of an undetected peroxide accumulation would be proportionally larger and correspondingly more dangerous.

There is a particular, almost paradoxical safety challenge embedded in peroxide forming chemicals that deserves explicit acknowledgment, since it complicates what might otherwise seem like a straightforward disposal solution to an aging, potentially hazardous container. Once a container of peroxide forming solvent has been allowed to sit long enough, and under conditions permissive enough, to accumulate a genuinely dangerous peroxide concentration, the safest course of action is often not to simply discard it through ordinary channels, since the mechanical disturbance involved in moving, transporting, and processing waste containers is precisely the kind of shock capable of triggering detonation in a sufficiently concentrated sample. This is why hazardous materials response guidance for confirmed or suspected high peroxide containers generally recommends against any unnecessary handling at all, instead calling for trained bomb disposal or hazardous materials personnel to assess and, where necessary, safely neutralize the material in place, treating an old bottle of laboratory solvent, in the most severe documented cases, with genuinely the same level of caution and specialized handling that would be applied to a conventional explosive device, a level of institutional seriousness that can feel, to someone encountering the guidance for the first time, wildly disproportionate to what appears, from the outside, to be nothing more than an unremarkable bottle of forgotten chemistry supplies.

The specific danger of academic research laboratories, as distinct from industrial or commercial chemical storage settings, deserves its own closing consideration in this chapter, since university research environments combine several of this book's recurring risk factors in a particularly concentrated form: comparatively inexperienced personnel, frequently graduate students working with hazardous materials for the first time, high staff turnover as students graduate and are replaced by new arrivals with no institutional memory of prior close calls, and, in many research settings, an experimental, exploratory culture in which established safety procedures can come to feel like a formality standing between a researcher and the specific scientific question they are actually trying to answer. A particularly well documented illustration of this dynamic occurred on January 7, 2010, in a chemistry laboratory at Texas Tech University, where a fifth year graduate student, working under a Department of Homeland Security funded research contract studying energetic, explosive compounds, was synthesizing a derivative of a compound called nickel hydrazine perchlorate, a material that, like the peroxide forming solvents examined earlier in this chapter, carries genuine shock and friction sensitivity once produced in sufficient quantity. The student had verbally been given a stated safety limit of one hundred milligrams for any single batch of the material, but subsequent investigation by the United States Chemical Safety and Hazard Investigation Board found that this limit had never been formally documented or consistently communicated across the research group, and on the day of the accident, the student and a first year student he was training had synthesized a batch weighing approximately ten grams, roughly one hundred times the informally understood safety limit. Believing, incorrectly, that wetting the compound with a solvent would render it safe to handle, the student began grinding the material with a mortar and pestle to prepare it for analysis, without wearing safety goggles at the time. The compound detonated in his hands, causing him to lose three fingers and suffer a serious eye injury along with burns and lacerations across his body.

The CSB's subsequent investigation into the Texas Tech accident reached conclusions that, by this point in the book, will read as a familiar and recurring theme, applied here to an academic rather than an industrial or agricultural setting. Investigators found that the university's existing chemical hygiene plan, while formally compliant with federal occupational safety standards, did not specifically require training on the physical, as opposed to purely toxicological, hazards of the materials students were working with, a distinction that mattered enormously given that the injury in this case arose from shock sensitivity and detonation risk rather than from any chemical toxicity. Investigators further found that the research group had experienced at least one prior near miss involving an unplanned and undocumented scale up of the same general synthesis, an early warning sign that, had it been formally logged and reviewed rather than passing unrecorded, might have prompted exactly the kind of tightened batch size controls and mandatory protective equipment requirements that could have prevented the January 2010 accident from occurring at all. The CSB's report, drawing on a widely used framework in accident investigation known as the Swiss cheese model, in which a serious accident generally requires multiple independent layers of an organization's safety system to fail simultaneously rather than any single point of failure, identified breakdowns at nearly every level of the university's safety culture, from the supervising faculty member's informal and inconsistently enforced batch size limit, up through the institution's broader absence of any systematic mechanism for capturing and disseminating lessons learned from near miss incidents across its many research laboratories.

The Texas Tech case, precisely because it involves university researchers rather than industrial workers, and a comparatively obscure energetic compound rather than a widely used industrial solvent, illustrates just how far this book's central argument extends beyond the large scale, headline disasters that dominate its other chapters. The same fundamental gap this book has traced repeatedly, between chemistry that is scientifically well understood and safety practice that has not been rigorously and consistently applied to the specific situation at hand, operates identically whether the setting is a Texas fertilizer warehouse holding thousands of tons of ammonium nitrate or a university laboratory bench holding a ten gram sample of an obscure energetic compound synthesized as part of a security research contract. The scale of consequence differs enormously between these settings. The underlying pattern of institutional failure, an informally communicated safety limit, an undocumented near miss, insufficient hazard specific training, does not.

Chapter 12

Reading the Warning Signs: Detection, Calorimetry, and the Science of Prevention

Every disaster examined in this book so far shares an uncomfortable common feature: in nearly every case, the hazard was, in principle, detectable in advance. Ammonium nitrate's capacity for violent decomposition under confinement and heat was documented chemistry, not a surprise, by the time West, Texas exploded in 2013. The specific reactive hazard that destroyed T2 Laboratories in 2007 was rooted in kinetics first formalized by Arrhenius in the 1880s. Diethyl ether's tendency to form shock sensitive peroxides has been documented in laboratory safety literature for generations. If the underlying chemistry was knowable, and in most cases had already been formally studied and published well before the disasters that make up this book's case studies, the obvious question becomes practical rather than scientific: what tools exist to actually detect a developing hazard before it escalates, and why, given that such tools exist and have existed for decades, do catastrophic exothermic accidents keep happening anyway. This chapter is about the science of prevention, the specific instruments and methods the chemical industry has developed to measure exothermic hazards in advance, and about the more sobering reality that measurement alone, however sophisticated, is never quite sufficient on its own.

The foundational tool in this domain is a laboratory technique called differential scanning calorimetry, generally abbreviated DSC, and its basic operating principle, though the instrumentation involved is considerably more sophisticated, would be recognizable in spirit to Antoine Lavoisier working with his ice calorimeter in the 1780s. A small sample of the material under investigation, perhaps a few milligrams of a chemical intermediate a company is considering using in a new industrial process, is placed in a small pan alongside a second, empty reference pan, and both pans are heated according to an identical, precisely controlled temperature program. Because the two pans are heated identically, any difference in the amount of heat flow required to maintain that identical heating rate between the sample and the empty reference pan reveals something happening chemically within the sample itself, a melting point, a crystallization event, or, most relevant to this book, the onset of an exothermic decomposition reaction, visible on the resulting graph as a distinct peak in the heat flow at a specific, measurable temperature. DSC testing is now governed by standardized methods, including the American Society for Testing and Materials standard E537, specifically covering the thermal stability testing of chemicals by this technique, and it has become one of the standard first screening tools chemical engineers use when evaluating whether a new material or process carries latent thermal hazard. Industry guidance developed for interpreting DSC results includes specific numerical thresholds that function almost like an early warning classification system: a measured decomposition energy release above roughly eight hundred joules per gram, for instance, is generally treated as a signal of potential explosive hazard warranting considerably more careful follow up investigation before the material is used at any meaningful industrial scale.

DSC's great practical virtue, the ability to screen a large number of candidate materials quickly using only a small sample quantity, is also its principal limitation, because the technique's rapid, controlled heating rate does not accurately replicate the actual thermal conditions a chemical experiences inside a full scale industrial reactor, where heat generated by a slowly developing reaction may have no ready path to escape at all, an important distinction the field refers to as adiabatic versus non adiabatic conditions. This gap is precisely what a second, considerably more elaborate technique, accelerating rate calorimetry, or ARC, was developed to address. Where DSC heats a tiny sample rapidly and measures the resulting heat flow, an ARC instrument does something closer to the opposite: it takes a somewhat larger sample, places it inside a sealed and heavily insulated calorimeter bomb, and then actively adjusts the surrounding oven temperature throughout the test to match whatever temperature the sample itself reaches, specifically to prevent any heat from escaping the sample into its surroundings, a condition engineers call pseudo adiabatic, since it approximates, as closely as practical laboratory instrumentation allows, the same heat trapping conditions that exist inside a real industrial reactor lacking adequate cooling. Under these carefully engineered conditions, the ARC essentially recreates, in miniature and under controlled laboratory observation, the exact thermal runaway dynamic responsible for the T2 Laboratories explosion, allowing engineers to observe and precisely measure the onset temperature at which a given chemical begins self heating, the rate at which that self heating accelerates as temperature climbs, and the maximum temperature and pressure the runaway reaction ultimately reaches, all before that same chemical is ever produced or handled at a scale where an actual runaway would put lives and property at risk.

The information an ARC test produces feeds directly into some of the most consequential engineering decisions a chemical facility makes, decisions that connect this chapter's laboratory science back to the specific industrial disasters examined earlier in this book. Knowing the precise onset temperature at which a material begins to self heat allows engineers to establish a maximum safe process and storage temperature with a documented, defensible safety margin built in, rather than relying on prior operating experience alone, the same kind of experience that had run successfully for one hundred seventy four batches at T2 Laboratories before catastrophically failing on the one hundred seventy fifth. Knowing the rate at which a runaway reaction accelerates and the maximum pressure it can generate allows engineers to properly size emergency pressure relief systems, the safety valves and rupture discs intended to vent a reactor's contents before the vessel itself fails catastrophically, addressing directly the specific technical shortfall investigators identified at the Seveso ICMESA plant in 1976, where the installed bursting disc had been calibrated to a pressure threshold high enough that by the time it actually activated, the runaway reaction inside the reactor had already progressed far enough to generate substantial quantities of toxic dioxin as an unwanted decomposition byproduct.

This calorimetric science has, in recent years, found an entirely new and rapidly growing application well beyond its traditional home in bulk industrial chemical manufacturing, one that connects directly to the subject of an earlier chapter in this book. As lithium ion battery adoption has accelerated across consumer electronics, electric vehicles, and grid scale energy storage, battery researchers and manufacturers have increasingly turned to ARC testing specifically to characterize the thermal runaway behavior of individual battery cells and full battery packs, using essentially the same fundamental measurement principle originally developed for bulk industrial chemicals to instead determine the precise temperature at which a given battery cell design begins self heating, how rapidly that self heating accelerates, and crucially, whether and how readily thermal runaway in one cell propagates its heat to trigger runaway in physically adjacent cells within the same pack, a specific and active area of contemporary ARC instrumentation development given how directly it feeds into battery pack safety engineering. The same laboratory technique that helps a chemical company determine the safe operating temperature ceiling for an industrial reactor now also helps a battery manufacturer determine how much physical separation and thermal barrier material needs to sit between individual cells in a consumer electric vehicle battery pack to prevent a single cell's failure from cascading into a fire consuming the entire vehicle.

Beyond calorimetry specifically, the broader engineering discipline that has grown up around exothermic hazard prevention operates under a guiding philosophy generally referred to as inherent safety, a concept that represents a meaningful and relatively modern shift in how the chemical industry approaches hazard management. Rather than treating hazard prevention primarily as a matter of adding external protective layers, alarms, relief valves, emergency shutdown systems, and operator training, all valuable and all still very much in active use, inherent safety asks a logically prior question: can the underlying process itself be redesigned so that the hazard is reduced or eliminated at its source, rather than merely contained after the fact. Industry literature on inherent safety design generally organizes this approach around four core strategies. Intensification involves minimizing the actual quantity of a hazardous substance present in a process at any given moment, for instance by reducing the physical size of a reactor or the length of piping through which a dangerous intermediate must travel, on the straightforward logic that a smaller inventory of hazardous material produces a smaller worst case consequence if something does go wrong. Attenuation involves using a hazardous substance under less severe conditions than would otherwise be required, lower temperature, lower pressure, or a more dilute concentration, wherever the underlying chemistry allows it. Simplification involves designing a process with fewer steps, fewer transfer points, and fewer opportunities for human error or equipment failure to introduce an unplanned deviation. And substitution involves replacing a particularly hazardous material or reaction pathway with a less hazardous alternative capable of achieving a comparable end result, a strategy directly relevant to the ongoing shift in the battery industry toward more thermally stable lithium iron phosphate chemistries discussed in an earlier chapter, chosen specifically because they are inherently, chemically less prone to runaway than some of the higher energy density alternatives they are gradually displacing.

One particularly striking illustration of the intensification principle in action involves microreactor technology, in which chemical reactions that would traditionally be run in large batch vessels, the same basic reactor design responsible for the T2 Laboratories disaster, are instead conducted continuously within networks of very small, precisely engineered channels, sometimes only millimeters in diameter. Because so little reactive material is present within the reactor system at any given instant, and because the enormous surface area to volume ratio of these small channels allows heat generated by an exothermic reaction to be removed far more efficiently than is possible in a large batch vessel, industry research has found that microreactor designs can substantially improve the inherent safety of strongly exothermic reaction processes specifically, offering a genuinely different engineering approach to the same underlying thermal runaway hazard rather than simply adding another layer of monitoring and alarm equipment around a fundamentally unchanged, large scale batch process.

None of this technology, however sophisticated, changes the sobering pattern this book has documented across its industrial case studies, and it is worth closing this chapter with that pattern stated plainly rather than allowing the genuine, real engineering progress described above to imply a false sense of resolved risk. The calorimetric tools described in this chapter, DSC and ARC testing, have existed in something close to their modern form for decades, well before several of the disasters examined earlier in this book occurred. Reactive hazard testing was not some emerging science T2 Laboratories lacked access to in 2007. It was an established discipline the company's own chemists and engineers evidently did not fully apply to their own process, a gap the CSB's investigation traced back to specific, identifiable shortfalls in chemical engineering education rather than to any absence of available testing technology. This is, in the end, the central and somewhat uncomfortable lesson this chapter has to offer: the science of predicting and preventing exothermic disaster is genuinely excellent, genuinely mature, and genuinely capable, when properly applied, of identifying a dangerous reaction long before it reaches the point of no return. Whether that science actually gets applied, consistently and rigorously, at every facility handling every hazardous material, remains, as it has been throughout the history recounted in this book, a question of institutional discipline and resource allocation rather than a question of what is scientifically possible.

Beyond the specific instruments and design philosophies described above, the broader discipline of process safety management has developed a formal analytical framework specifically intended to close the gap between what calorimetric testing reveals about a chemical's inherent hazard and what actually happens on the plant floor, a framework known as hazard and operability study, commonly abbreviated HAZOP, which originated within the chemical industry in the 1960s and has since become a standard, and in many jurisdictions a legally required, component of process safety review for facilities handling hazardous materials. A HAZOP review brings together a multidisciplinary team, typically including process engineers, operators with direct hands on experience running the equipment under review, and safety specialists, to systematically examine every stage of a chemical process using a structured set of guide words, more, less, none, reverse, and other, applied methodically against every relevant process parameter, temperature, pressure, flow rate, concentration, in order to surface every plausible deviation from intended operating conditions and to assess, deliberately and in advance, what the consequences of that deviation might be and whether existing safeguards are genuinely adequate to prevent or contain it. Applied rigorously to a process like the one operating at T2 Laboratories in 2007, a HAZOP review asking a structured question along the lines of what happens if cooling is lost during the metalation step would have forced exactly the kind of explicit, documented engagement with worst case reactive hazard consequences that the CSB's subsequent investigation found had never adequately occurred at the company, a sobering illustration of how a formal analytical procedure, applied consistently rather than only after a disaster has already provided the motivation, can surface precisely the kind of latent hazard that a purely informal, experience based approach to process safety, however well intentioned, can allow to go unrecognized for years.

The specific value of redundant, independently verified instrumentation, one of the central engineering principles a rigorous HAZOP review is meant to enforce, is illustrated with unusual clarity by an incident that occurred not inside a reactor vessel at all but within an ordinary fuel storage tank, on the early morning of December 11, 2005, at the Buncefield oil storage depot near Hemel Hempstead, in Hertfordshire, England. Buncefield was, at the time, one of the largest fuel storage facilities in the United Kingdom, and on the night before the disaster, one of its storage tanks, designated Tank 912, was being filled with unleaded petrol as part of an entirely routine overnight operation. The tank's safety design relied on two independent and, critically, supposedly redundant systems to prevent overfilling: a standard gauge that allowed control room staff to monitor the fuel level directly, and a wholly separate independent high level switch, engineered specifically to automatically halt the filling operation if the tank's contents ever approached the top of the vessel regardless of what the primary gauge indicated, a deliberate second layer of protection built on the sound engineering principle that no single instrument should ever be trusted as the only barrier standing between a routine operation and a catastrophic failure.

On this particular night, both systems failed simultaneously, through entirely unrelated mechanisms that happened to converge with devastating effect. The primary gauge had a documented history of intermittently sticking since a servicing operation earlier that year, a known defect that, on this occasion, caused it to display a stalled, artificially low reading even as the tank continued filling well past its intended capacity. The independent high level switch, the dedicated backup specifically designed to catch exactly this kind of primary gauge failure, proved unable to perform its function for reasons that, upon investigation, turned out to be almost absurdly mundane: the switch required a padlock to hold its mechanical check lever in an active, working position, and that padlock had not been fitted, leaving the supposedly independent safety backup effectively disengaged and non functional throughout the entire overnight filling operation, without any alarm or indication to control room staff that their second layer of protection was, in practice, providing no protection at all. Over the course of roughly forty minutes, more than three hundred tons of petrol overflowed from the tank's roof vents, and because the night was unusually still and windless, the resulting vapor rather than dispersing harmlessly into the atmosphere instead pooled and spread across the site, eventually forming a dense vapor cloud investigators later estimated at some three hundred sixty meters in diameter before it found an ignition source and detonated in what has been described as the largest explosion in peacetime Europe in half a century, an event powerful enough to register 2.4 on the Richter scale and to be picked up by seismograph stations as far away as the Netherlands.

The explosion and the prolonged fire that followed, which burned for several days and produced a column of black smoke visible across much of southern England, injured more than forty people and caused extensive damage to homes and businesses well beyond the depot's own boundary, and remarkably, given the scale of destruction, resulted in no fatalities, a fact investigators and safety analysts have generally attributed to the disaster's timing in the early hours of a quiet Sunday morning, when the industrial areas surrounding the depot were largely unoccupied, a circumstance of pure timing rather than of any safety measure functioning as intended. The subsequent investigation, conducted under the United Kingdom's Control of Major Accident Hazards regulations, identified the simultaneous, independent failure of both the primary gauge and the backup high level switch as the direct technical cause, but its findings extended well beyond that immediate mechanical failure to broader and more systemic conclusions regarding what investigators termed secondary and tertiary containment, the additional physical barriers, bunds, drainage systems, and containment walls that are meant to limit the consequences of a spill even after primary containment has already failed, layers of protection that Buncefield's own design proved considerably less effective at containing a large scale vapor release than had been assumed.

Buncefield's inclusion in this chapter, alongside the calorimetric and hazard analysis techniques described above, illustrates a point about detection and prevention science that deserves to be stated plainly: sophisticated instrumentation and formally redundant safety systems are only as reliable as their weakest actual point of maintenance and verification, and a safety system that exists on an engineering drawing but has been quietly allowed to degrade in practice, whether through a sticking gauge nobody had gotten around to properly repairing or a padlock nobody had thought to check, provides no more real protection than if it had never been installed at all. The specific failure at Buncefield, a redundant safety switch rendered useless by a missing padlock, is almost startling in its ordinariness, and it belongs, in that sense, alongside several of the other disasters examined throughout this book, a reminder that the gap between a well designed safety system on paper and a genuinely functioning one in practice is very often bridged, or left dangerously unbridged, by exactly this kind of small, unglamorous maintenance detail that no amount of sophisticated calorimetry or process design can substitute for.

The layered defense in depth philosophy that underlies both calorimetric testing and structured hazard review procedures like HAZOP reflects a broader shift in how the modern chemical and process industries conceptualize safety, one worth naming explicitly as this chapter draws toward its conclusion. Older approaches to industrial safety tended to treat accident prevention primarily as a matter of protecting workers from hazards that were essentially taken as fixed, mandating protective equipment, establishing safe operating procedures, and installing emergency response systems around a process whose underlying hazard profile was accepted as a given constraint. The inherent safety and structured hazard analysis approaches described in this chapter represent a meaningfully different philosophy, one that asks whether the hazard itself can be reduced, minimized, or eliminated at the design stage, before any worker ever needs to be protected from it at all. This shift, still incomplete and unevenly adopted across different industries, regions, and companies of varying size and sophistication, represents nonetheless one of the genuinely hopeful developments running through the otherwise sobering history this book has recounted, a recognition, gradually if unevenly translating into practice across the decades since Seveso and T2 Laboratories, that the most effective response to an exothermic hazard is very often not a better way of surviving its consequences, but a redesigned process that never allows those consequences to become possible in the first place.

Chapter 13

The Missing Memo: Piper Alpha, Texas City, and the Human Side of Combustion

Not every case study in this book involves a reaction that was, in a strict sense, poorly understood or newly encountered. Some of the worst disasters in industrial history occurred not because anyone misjudged the underlying combustion chemistry, hydrocarbon fuel burning in the presence of oxygen and an ignition source is about as thoroughly understood a reaction as exists in all of applied chemistry, but because a shift change, a missing piece of paperwork, and roughly ten minutes of avoidable confusion combined to hand that entirely familiar chemistry an opportunity it should never have been given. The Piper Alpha disaster, which destroyed a North Sea oil and gas platform on the night of July 6, 1988, and killed one hundred sixty seven of the two hundred twenty six people aboard, remains the deadliest offshore oil disaster in history, and its root cause offers this book's clearest illustration yet of a theme introduced in earlier chapters: that the chemistry of exothermic disaster is very often the easy part to understand, while the organizational and communication failures that let that chemistry loose are considerably harder to engineer against.

Piper Alpha, operated by Occidental Petroleum roughly one hundred twenty miles northeast of Aberdeen, Scotland, had been commissioned in 1976 as an oil production platform and was subsequently modified in 1978 to also process natural gas gathered from two neighboring platforms, Tartan and Claymore, a modification that meant the facility was, by the night of the disaster, handling substantial volumes of high pressure gas in addition to its original oil production role, all within a physical structure whose safety systems, as investigators later determined, had never been fully redesigned to reflect that expanded and considerably more hazardous mission. On the morning of July 6, routine maintenance began on a backup condensate pump, designated Pump A, which required removing its pressure safety valve for inspection and overhaul, a piece of equipment specifically intended to relieve dangerous overpressure and one whose absence, even temporarily, transformed Pump A from a normal piece of process equipment into a genuine hazard if anyone attempted to operate it before the valve was restored. Standard procedure called for this kind of in progress maintenance to be logged clearly enough that oncoming shift workers would understand, without any ambiguity, that the affected equipment could not be safely started.

That evening, the platform's other condensate pump, Pump B, unexpectedly failed, and with gas production needing to continue, control room staff searched for the paperwork that would tell them the status of Pump A, the obvious alternative to bring online in Pump B's place. According to the subsequent public inquiry, chaired by Lord William Cullen, the written notification documenting that Pump A's safety valve had been removed and that the pump was consequently unsafe to operate was not found by the staff on duty that night, a breakdown in what safety engineers now call shift handover communication, the formal and informal transfer of critical operational knowledge from one working shift to the next. Unaware that the safety valve had been removed and that the resulting gap in the pipework had been sealed only with a temporary blind flange, control room staff restarted Pump A at approximately 9:55 in the evening. Within minutes, high pressure hydrocarbon gas began leaking through that inadequately secured temporary seal, and at approximately 10:00 p.m., the resulting vapor cloud found an ignition source and exploded.

What followed, over the course of roughly twenty two minutes according to subsequent reconstruction, was a cascading sequence of failures that transformed a serious but potentially containable initial explosion into a total catastrophe, and each stage of that cascade illustrates a distinct engineering or procedural principle this book has examined in other contexts. The initial blast blew through the firewalls separating the platform's various operational modules, and investigators later identified a critical design flaw directly relevant to this book's broader argument about engineering for the specific hazard actually present rather than a generic or outdated one: because Piper Alpha had originally been designed and built as an oil only platform, its firewalls had been engineered to resist fire, a slower, more containable threat, but not to withstand the sudden overpressure of an explosion, a design mismatch that had never been corrected even after the platform's 1978 conversion to also handle high pressure gas, a considerably more explosion prone hazard. The explosion ruptured additional oil and condensate lines, feeding a rapidly escalating fire, and critically, the platform's automatic water deluge system, specifically designed to suppress exactly this kind of fire, had been switched to manual operation and was never activated, a widely reported detail from the subsequent inquiry attributed to a standing precaution that kept the system's water intake pumps offline whenever divers were working near them, a safety measure for one hazard that, on this occasion, removed protection against a considerably larger one.

Approximately twenty minutes after the initial explosion, the fire had grown hot enough to rupture the risers, the large diameter pipes carrying gas up from the seabed, connecting Piper Alpha to the neighboring Tartan and Claymore platforms, and because those neighboring platforms continued feeding gas into the shared pipeline system even as Piper Alpha burned, communication and coordination gaps between the connected platforms meant that a fresh and enormous supply of fuel, reported in some technical accounts to have reached a rate on the order of thirty tons of gas per second, continued arriving at the burning platform for a considerable period after the initial explosion, feeding a fireball of such scale that, at its peak intensity, the rate of natural gas being consumed by the fire was reportedly comparable to the entire United Kingdom's natural gas consumption at that moment, an almost unimaginable figure that illustrates just how much stored chemical energy a well developed offshore gas processing and gathering hub represents when its containment fails completely.

The human toll of the disaster reflected the platform's specific design as much as the fire itself. Many crew members, following standard emergency protocol that called for gathering in a designated accommodation module to await helicopter evacuation, found themselves trapped there as the surrounding fires made both sea and air evacuation impossible, and subsequent analysis of the one hundred sixty seven deaths found that the majority resulted from smoke inhalation within that accommodation area rather than from burns or the explosions themselves, a finding that prompted, in the disaster's regulatory aftermath, a fundamental rethinking of offshore emergency evacuation procedures away from a strategy of centralized shelter in place toward greater emphasis on rapid, independent evacuation capability.

The Cullen Inquiry's findings, published in 1990, identified failures at essentially every level of the organization operating Piper Alpha, from the specific, immediate breakdown in shift handover communication that allowed Pump A to be restarted without anyone realizing its safety valve was missing, up through broader and more systemic findings regarding permit to work systems, the formal documentation and sign off procedures meant to ensure that maintenance work on hazardous equipment is properly tracked and communicated across shifts, which the inquiry found had been allowed to degrade into a largely procedural formality rather than the rigorous safety control it was designed to be. This finding proved enormously consequential for offshore safety regulation well beyond the specific platform involved, prompting the United Kingdom to fundamentally restructure its offshore safety regulatory framework, transferring oversight to a dedicated safety authority and mandating a formal safety case regime under which operators must proactively demonstrate, to a regulator's satisfaction, that major hazards have been identified and adequately controlled, rather than simply complying with a checklist of prescriptive technical rules, a shift in regulatory philosophy that has since influenced process safety regulation well beyond the offshore oil and gas sector specifically.

Piper Alpha's inclusion in this book alongside a second, geographically and organizationally distinct disaster helps demonstrate that its central lesson, hydrocarbon combustion chemistry proving fully understood even as the human systems managing it fail, was not a one time or isolated organizational failure. On March 23, 2005, at approximately 1:20 in the afternoon, a series of explosions tore through the isomerization unit at BP's refinery in Texas City, Texas, a process unit specifically designed to boost the octane rating of gasoline components, killing fifteen contract workers and injuring roughly one hundred eighty more, nearly all of them working out of temporary office trailers that had, in a decision the subsequent investigation identified as a critical and avoidable error, been positioned in close proximity to process equipment known to carry a significant hazard. The disaster's immediate technical cause traces to a startup procedure for a distillation column called the raffinate splitter, which was being brought back online following a maintenance outage. Operators, working with what a subsequent United States Chemical Safety and Hazard Investigation Board report identified as poor level monitoring and malfunctioning instrumentation, allowed the column to overfill with liquid hydrocarbon far beyond its safe operating capacity, and as pressure built inside the overfilled vessel, the unit's pressure relief system activated exactly as designed, but vented directly into a blowdown stack that, critically, had never been equipped with a flare system capable of safely burning off the released hydrocarbon vapor, a design gap that allowed a geyser of flammable liquid to erupt directly into the open air above the unit. That vapor cloud found an ignition source, by most subsequent accounts the idling engine of a nearby diesel truck, and the resulting explosion and fire killed the workers in the trailers positioned nearby before they had any meaningful opportunity to evacuate.

The Chemical Safety Board's investigation, and a separate, independent review commissioned by BP itself and led by former United States Secretary of State James Baker, together identified a failure pattern that reads, point for point, as an American echo of the organizational findings from the Cullen Inquiry into Piper Alpha nearly two decades earlier. Investigators found deviations from standard operating procedures during the unit's startup, malfunctioning equipment that maintenance records had been closed out as repaired without the underlying repair work actually having been completed, and, in a detail that connects directly back to this chapter's account of the missing paperwork at Piper Alpha, a breakdown in communication during shift handover that left incoming operators without a full and accurate picture of the unit's actual operating status. The Baker Panel's broader findings extended well beyond this single unit and this single incident, identifying what investigators characterized as deeper cultural and organizational deficiencies in how the company approached process safety across its refining operations generally, including a tendency to treat personal safety metrics, individual injury and incident rates, as an adequate proxy for the entirely distinct question of process safety, whether the underlying chemical and mechanical systems themselves were being adequately managed to prevent a catastrophic loss of containment, a conflation the investigation identified as a significant contributing factor to leadership's apparent confidence in the refinery's safety performance right up until the moment fifteen people died.

Reading Texas City alongside Piper Alpha, separated by seventeen years and by an ocean, but connected by the same underlying pattern, well understood hydrocarbon combustion chemistry, a pressure relief and containment system with a specific, identifiable design gap, and a shift handover and procedural culture that failed to catch a developing hazard before it became irreversible, reinforces this chapter's central argument with an additional and considerably more recent data point. Process safety as a distinct engineering and organizational discipline, separate from the more general and more commonly tracked metrics of personal workplace safety, has continued to mature significantly in the decades since Piper Alpha first prompted its formal articulation, and the Baker Panel's specific finding regarding the danger of conflating the two categories of safety has itself become a standard cautionary reference across the process safety literature in the years since. Yet Texas City demonstrates that the specific human and organizational vulnerabilities Piper Alpha first brought to such devastating public attention, imperfect shift communication, maintenance records that do not reliably reflect the true state of critical equipment, and relief systems whose design has not kept pace with the actual hazard they are meant to manage, remain live and recurring risks even at some of the world's largest, best resourced, and most technically sophisticated industrial operators, a sobering reminder that organizational learning, unlike the chemistry it is meant to help manage, does not stay learned automatically, and must instead be actively and continuously maintained across the same kind of institutional memory this book has found wanting, again and again, throughout its many other case studies.

Chapter 14

Living With Exothermic Risk: What History Teaches Us About Respect, Not Fear

There is a reason this book has moved back and forth so often between spectacular, sudden industrial catastrophes and quiet, slow, easily overlooked domestic hazards, and that reason has less to do with chemistry than with the workings of the human mind. Decades of research in the psychology of risk perception, much of it originating with the psychologist Paul Slovic and collaborators working in what is now called the psychometric paradigm, has established a finding that this book's case studies illustrate again and again without ever naming directly until now: human beings are remarkably poor at intuitively weighing risk according to its actual statistical likelihood or consequence, and remarkably consistent in the specific, predictable ways that intuition goes wrong. Understanding those specific ways is not a digression from the chemistry examined in the preceding chapters. It is, in a real sense, the missing piece needed to explain why so many of this book's disasters happened at all, given that the underlying science was, in nearly every case, already well understood by someone, somewhere, before the reaction in question ran away.

The psychometric research identifies two dimensions that, more than almost any other factor, drive how frightening a given hazard feels to an ordinary person, independent of that hazard's actual statistical danger. The first is what researchers call dread, a cluster of characteristics including a perceived lack of personal control over the hazard, catastrophic potential, involuntary exposure, and consequences that feel severe and irreversible. The second is what researchers call the unknown risk dimension, encompassing how unfamiliar, how poorly understood, and how difficult to directly observe a hazard is, along with whether its consequences tend to be delayed rather than immediate. Hazards that score high on both dimensions, unfamiliar and dreadful at once, tend to provoke intense public concern regardless of their actual statistical frequency, while hazards that score low on both, familiar and seemingly ordinary, tend to be substantially underestimated even when their actual, cumulative toll is considerably larger. Nuclear power, for reasons that have been extensively studied and that connect directly to this framework, consistently ranks among the most dreaded hazards in psychometric surveys, generating intense public anxiety even in the absence of any nearby incident, while car accidents, a genuinely enormous and well documented source of injury and death worldwide, are processed by most people with something close to equanimity, largely because driving feels familiar, feels voluntary, and feels, whether or not this feeling is statistically justified, controllable.

This asymmetry maps almost perfectly onto the structure of this book. A researcher studying dread risk directly compared what psychologists term dread risks, rare, sudden events that kill a large number of people at once, such as a terrorist attack or an earthquake, against what they term continuous risks, comparatively frequent events that kill an equally large number of people spread out gradually over a longer period, such as diabetes or ordinary traffic accidents, and found that people react far more strongly, in terms of both fear and active avoidance behavior, to the sudden, concentrated version of harm even when the two scenarios produce statistically identical total losses. The well documented pattern following the September 11, 2001 terrorist attacks illustrates this dynamic with particular clarity and particularly grim statistical consequence: a substantial number of Americans avoided air travel in the months following the attacks and shifted to driving instead, a choice that felt, in the moment, like a rational response to a dreaded and unfamiliar threat, but one that researchers have since calculated actually increased the total number of fatal highway crashes during that period, since driving carries a meaningfully higher statistical risk of death per mile traveled than commercial air travel, dread risk or otherwise. People were not behaving irrationally in any simple sense. They were behaving exactly as the psychometric research predicts humans behave, weighting a dramatic, concentrated, unfamiliar risk far more heavily than a familiar, gradual, statistically larger one.

Every chapter in this book can be read as a variation on this same asymmetry, translated into the specific vocabulary of exothermic chemistry. The West, Texas fertilizer explosion and the Bhopal disaster are, in the language of the psychometric paradigm, close to textbook dread risks: sudden, catastrophic, largely outside the direct control of the people affected, and consequently the kind of event that dominates news coverage for weeks and prompts genuine, sweeping regulatory response, exactly the pattern that followed Bhopal, Seveso, and, to a somewhat lesser but still meaningful degree, West, Texas. A hay barn slowly self heating over six weeks, a laboratory bottle of ether quietly forming peroxides over a decade of storage, or a compost pile in someone's backyard generating heat through ordinary microbial activity are, by contrast, close to the opposite profile: gradual, familiar, embedded in ordinary agricultural or domestic routine, and largely invisible until the moment they are not, which is precisely why this book has argued throughout that these slower, quieter hazards are so persistently and dangerously underestimated relative to their actual cumulative toll. Farmers lose more barns to spontaneous hay combustion than to lightning strikes, year after year, in part because a lightning strike fits neatly into an existing mental category, sudden bad luck striking from outside, while a hay fire's true cause, the farm's own stored feed quietly cooking itself from within, resists that same intuitive framing.

This same asymmetry helps explain a pattern in household risk that connects the grease fire chapter directly to this psychological framework. Cooking is, by a wide statistical margin, the leading cause of home fires in the United States, a familiar, voluntary, everyday activity that the vast majority of people perform without a second thought regarding its risk profile, precisely the combination of familiarity and perceived controllability the psychometric research identifies as systematically associated with underestimated danger. A person who would never dream of storing gasoline in an open container inside their kitchen, correctly recognizing gasoline as an obviously hazardous substance, will nonetheless leave a pan of oil heating unattended on a stovetop for several minutes without a moment's concern, even though, as detailed earlier in this book, that same pan of oil is entirely capable of reaching its autoignition temperature and combusting spontaneously within as little as eighteen minutes, no external spark required. The oil does not feel dangerous the way stored gasoline feels dangerous, not because it is meaningfully less capable of causing serious harm, but because cooking itself is such a deeply familiar, voluntary, and seemingly controllable activity that the specific thermal chemistry taking place inside the pan rarely registers as a distinct category of hazard at all.

None of this is presented as a criticism of ordinary human judgment, and it would be a mistake to read this book's closing chapter as an argument that people should somehow learn to feel equally afraid of everything in proportion to its precise statistical risk, an aspiration that runs directly counter to how human cognition has ever actually worked and that would, in any case, likely produce a kind of generalized anxiety considerably less useful than the specific, targeted literacy this book has aimed to build instead. The goal of understanding exothermic hazards is not to increase fear but to redirect attention, and attention, unlike fear, is a resource that can be deliberately and usefully allocated once a person understands where the actual chemistry warrants it. A farmer who understands the moisture threshold and the fourth to sixth week risk window for spontaneous hay combustion does not need to feel constant dread about the barn. They need only check the internal temperature of a new hay stack with a probe thermometer during that specific window, a five minute task performed at the right time rather than a source of ongoing anxiety. A home cook who understands why water and burning oil are catastrophic together does not need to fear cooking. They need only know, in the specific moment a grease fire actually starts, to reach for a lid or a fire blanket rather than the tap, a piece of specific, actionable knowledge that requires no ongoing vigilance at all, only correct information available at the single moment it matters.

A second, closely related cognitive pattern identified by the psychologists Amos Tversky and Daniel Kahneman in the early 1970s, working independently of but in close intellectual company with Slovic's psychometric research, offers a further and genuinely useful piece of this puzzle, and it concerns not what makes a risk feel frightening but what makes a risk feel probable in the first place. Tversky and Kahneman termed this pattern the availability heuristic, the tendency to judge how likely or how common something is by how easily examples of it come to mind, rather than by any more rigorous accounting of its actual statistical frequency. This heuristic serves people reasonably well in a great many ordinary circumstances, since genuinely common events do tend, on average, to be easier to recall than genuinely rare ones, but it fails in specific and predictable ways whenever the ease of recalling something and its actual frequency come apart, which happens most often with events that are vivid, recent, or emotionally striking regardless of how common they truly are. A dramatic industrial explosion covered extensively across national news media becomes, for months or years afterward, unusually easy to recall, and consequently feels unusually likely to recur, while a slow, undramatic hazard, a hay stack quietly self heating in a barn, a bottle of solvent aging unnoticed on a laboratory shelf, generates no comparably vivid memory for the availability heuristic to seize upon, and consequently continues to feel, however incorrectly, like a background risk barely worth tracking.

This same dynamic operates within organizations as much as within individual minds, and it offers a further explanation for a pattern this book has traced repeatedly across its industrial case studies: the tendency for a specific, well documented hazard to fade from institutional attention in the years following the disaster that first brought it to light, only for a strikingly similar disaster to recur, sometimes decades later, once the vivid memory of the original event has faded from the minds of the people currently responsible for managing the hazard. Oppau in 1921 was, at the time, an enormous and internationally reported catastrophe, yet its lessons had evidently faded enough within the chemical industry by 1947 that a strikingly similar ammonium nitrate disaster occurred at Texas City, and faded again sufficiently by 2013 that West, Texas repeated the same underlying failure a third time, and again by 2020, when Beirut repeated it a fourth. This is not because the relevant safety literature disappeared or because the chemistry became less well understood with each passing decade. It is, at least in significant part, because the specific, vivid, personally salient memory of the previous disaster gradually faded from the working knowledge of whichever generation of workers, managers, and regulators happened to be making decisions at each subsequent facility, replaced by newer, more immediately available memories of entirely different and more recent concerns, a form of organizational forgetting that mirrors, at institutional scale, the same availability heuristic operating within any single person's mind.

This is, in the end, the practical philosophy this book has tried to model across its case studies rather than state as an abstract principle until now: exothermic risk is best managed not through generalized caution applied evenly across every material and situation, an approach that quickly becomes exhausting and ultimately self defeating, but through specific, well targeted knowledge applied at the specific points where the underlying chemistry actually creates hazard. Understanding delta H and activation energy, the concepts introduced in this book's first chapter, provides the conceptual foundation. Understanding the particular mechanism at work in each specific domain examined since, microbial handoff to chemical oxidation in a hay stack, Coulomb explosion in a droplet of reactive metal, autoxidation in an aging bottle of solvent, internal short circuit cascading through a lithium battery pack, provides the specific, transferable pattern recognition that allows a person to notice, correctly, when an ordinary looking situation actually warrants a closer look. None of the disasters chronicled in this book occurred because the relevant chemistry was unknowable. Each occurred because specific, available knowledge failed, in that particular moment, to reach the specific person who needed it in time to act on it. Closing that gap, one reader and one piece of accurate, well explained chemistry at a time, has been this book's entire purpose from its opening page.

It is worth addressing directly a question that an attentive reader may have been holding since early in this book: if the underlying chemistry of every hazard examined here has been well understood for decades, sometimes for the better part of two centuries, why does this book exist at all, rather than simply pointing readers toward the extensive existing technical and regulatory literature on each individual topic. The honest answer returns to the psychometric framework introduced earlier in this chapter, and to a specific limitation of technical and regulatory literature that this book has tried, deliberately, to address. A

Chemical Safety Board investigation report, however thorough and however scientifically rigorous, is written primarily for an audience of safety professionals, engineers, and regulators, in a register and at a level of technical density that, however appropriate for its intended purpose, rarely reaches the wider public whose daily lives intersect with exothermic hazards just as directly, if considerably less dramatically, than the workers at a chemical plant. A farmer checking hay temperature, a home cook reaching for a lid rather than water, a parent deciding where to charge a child's e scooter overnight, none of these people are likely to read a CSB investigation report or an academic paper on Arrhenius kinetics, and this book's specific purpose has been to translate the substance of that technical and regulatory literature into a form ordinary readers will actually encounter, actually finish, and actually remember at the specific moment its content becomes practically relevant.

This translation function matters more than it might initially appear, precisely because of the asymmetry in risk perception described earlier in this chapter. Formal regulation and professional safety literature tend to concentrate, appropriately given their institutional purpose, on exactly the high dread, high consequence hazards that already receive substantial public and governmental attention, industrial reactors, bulk chemical storage, nuclear facilities, the kinds of settings where a single catastrophic failure can kill dozens of people at once and consequently attracts sustained regulatory scrutiny in its aftermath. The quieter, more familiar hazards examined throughout much of this book, hay barns, kitchen stovetops, laboratory shelves, individual consumer batteries, generally receive proportionally less sustained public attention precisely because no single incident tends to be large enough to command the same institutional response, even though the cumulative toll across thousands of smaller, less publicized incidents, farm fires, kitchen fires, laboratory accidents, consumer battery failures, adds up, year after year, to a burden of injury and property loss that rivals or exceeds the toll of the more dramatic industrial catastrophes this book has also examined. Closing that attention gap, directing a reasonable and proportionate amount of informed attention toward the quieter hazards that ordinary daily life actually involves, has been as much a part of this book's purpose as recounting the more dramatic industrial history that opens several of its chapters.

The genuine, measurable engineering progress documented in the preceding chapter, improved calorimetric testing, inherently safer process design, more thermally stable battery chemistries, offers real and justified grounds for cautious optimism about the trajectory of this subject, even as the underlying thermodynamics that make exothermic reactions hazardous in the first place will never change, and never need to. Fire, in the broadest chemical sense this book has used the term, has been humanity's oldest and most valuable technology for almost the entire span of human history, and every hazard examined in these pages represents not some new and unprecedented danger but simply fire behaving according to rules that have applied, unbroken, since long before anyone was present to study them. What has changed, gradually and unevenly across the specific domains this book has traced, from nineteenth century thermochemistry to twenty first century battery calorimetry, is not the chemistry itself but the depth and specificity of human understanding of it, and the accumulating, if still incomplete, translation of that understanding into engineering, procedure, and the kind of ordinary, well informed caution that allows a genuinely dangerous process to be used safely, productively, and without unnecessary fear, by people who simply know, correctly, what it is that they are actually looking at.

Chapter 15

The Positive Feedback Problem: Chernobyl, Fukushima, and the Physics of Runaway

Not every runaway reaction examined in this book begins with a chemical bond. At 1:23 in the morning on April 26, 1986, at the Chernobyl Nuclear Power Plant near the town of Pripyat in what was then the Soviet Union, operators of the Unit 4 reactor were in the final stages of a planned safety test, one designed to determine whether the plant's steam turbines could continue generating enough electricity to power emergency cooling pumps during the brief window between a loss of external power and the startup of backup diesel generators. The test itself had nothing to do with exothermic chemistry in the sense this book has used the term throughout its preceding chapters. It concerned nuclear fission, an entirely different category of energy release, governed by physics rather than chemistry. And yet what happened over the following ninety seconds, as the reactor's power output surged catastrophically out of control, followed a pattern this book's readers will by now find distinctly, almost eerily familiar: a positive feedback loop in which a small deviation fed upon itself, accelerating faster than any available safety system could intervene, until the physical structure containing the reaction failed entirely.

The specific mechanism responsible is known as a positive void coefficient, and understanding it requires a brief detour into how the RBMK reactor design at Chernobyl actually worked, since the flaw in question was baked directly into the reactor's architecture rather than arising from any chemical instability in its fuel. The RBMK, a Soviet designed graphite moderated, water cooled reactor, used graphite blocks surrounding the fuel channels to slow, or moderate, the neutrons released by nuclear fission, a necessary step to sustain the fission chain reaction efficiently, while water flowing through separate channels served the dual purpose of cooling the fuel and carrying away steam to drive the plant's turbines. In most Western reactor designs, water serves as both moderator and coolant simultaneously, which creates an inherently self limiting safety characteristic: if the reactor overheats and the water begins boiling into steam, the loss of that same water reduces the moderating effect needed to sustain the fission reaction, and the reactor's power output falls automatically, a negative feedback loop that tends to shut a runaway reaction down rather than accelerate it. The RBMK's separation of moderator, the solid graphite, from coolant, the water passing through separate channels, eliminated this built in safety margin entirely, and worse, inverted it. As water within the RBMK's cooling channels heated and began forming steam voids, the graphite continued moderating neutrons exactly as before, undiminished, while the loss of water, which under certain reactor conditions was absorbing more neutrons than it was helping to moderate, actually increased the rate of fission taking place. More steam voids meant more reactivity. More reactivity meant more heat. More heat meant more steam voids. The reactor, under specific low power operating conditions, was capable of feeding its own runaway acceleration in a feedback loop with no natural braking mechanism built into the underlying physics, the nuclear equivalent of the thermal runaway this book has traced through hay stacks, industrial reactors, and lithium battery cells, except operating on a timescale of seconds rather than hours or weeks.

The test being conducted that night required operators to reduce the reactor to a low power state, a condition under which the RBMK's positive void coefficient became especially pronounced and especially dangerous, and a series of operational decisions, some driven by an unplanned delay in the test schedule and a subsequent buildup of a neutron absorbing byproduct of fission called xenon that made the reactor increasingly difficult to control at low power, led operators to withdraw far more of the reactor's control rods, the devices used to absorb neutrons and suppress the fission rate, than the plant's own operating procedures permitted, in an effort to compensate for the xenon buildup and keep the reactor running. This left the reactor in a configuration with dramatically reduced margin for error and, compounding the danger further, the control rods themselves carried a specific design flaw: their tips, the portion that would enter the reactor core first upon an emergency shutdown, were made of graphite rather than the neutron absorbing boron carbide that made up the rest of each rod's length, a design choice that meant the initial effect of inserting the rods during an emergency shutdown was to briefly displace water and increase reactivity, precisely the opposite of the intended safety response, before the boron carbide portion of the rods finally began absorbing neutrons and slowing the reaction. When operators, alarmed by an unexpected power surge as the test proceeded, pressed the emergency shutdown button, they triggered exactly this flawed sequence: a brief, sharp spike in reactivity as the graphite tipped rods entered the core, arriving at precisely the wrong moment to intervene in a reactor whose positive void coefficient was already driving an accelerating chain reaction.

The result was a power excursion of a scale and speed the reactor's designers had not adequately anticipated, and within seconds, the fuel began to fragment and overheat catastrophically, producing a violent steam explosion that blew the massive concrete and steel lid off the reactor core, followed shortly after by a second explosion. The force of these explosions exposed the reactor's ruptured core directly to the atmosphere, and the graphite that had served as the reactor's moderator, now exposed to air at extreme temperature, ignited, beginning a fire that burned for days and that investigators have identified as a major contributing factor to the disaster's ultimate radiological severity, since the burning graphite acted as a mechanism actively lofting radioactive material from the exposed core high into the atmosphere, where it dispersed across a substantial portion of Europe, first detected, notably, not by any announcement from Soviet authorities but by radiation monitoring equipment in Sweden, which registered the plume days before Moscow publicly acknowledged that an accident had occurred at all.

The RBMK reactor, unlike most Western nuclear reactor designs of the same era, had not been built with a robust, hardened containment structure specifically engineered to survive and contain the consequences of a severe core damage event, a design decision that international nuclear safety analysis conducted after the disaster identified as a critical contributing factor to the scale of the resulting radiological release, since even a catastrophic core failure of the kind that occurred at Chernobyl might have been substantially contained, and its public health consequences substantially reduced, by a containment structure of the type standard in most reactors built to Western regulatory requirements. The absence of such a structure meant that once the reactor vessel itself failed, there was, quite simply, no further physical barrier standing between the exposed, burning reactor core and the surrounding environment, a design gap that connects directly to the broader theme of engineered barriers and containment systems running throughout this book, from the pressure relief systems examined in the chapter on industrial reactor safety to the thermal propagation barriers increasingly built into modern lithium battery packs. In every domain this book has examined, the question of how well a system is engineered to contain and arrest a runaway reaction, once prevention has already failed, proves at least as consequential as the question of how well that runaway was prevented from beginning in the first place.

Subsequent international assessment of the Chernobyl disaster, conducted over the years and decades following the accident, has generally converged on a conclusion this book's readers will recognize as a close structural echo of the T2 Laboratories and Seveso investigations examined in an earlier chapter: the disaster resulted not from any single failure but from a combination of design deficiencies inherent to the reactor itself and operational decisions and procedural violations made by plant staff attempting to complete a scheduled test under difficult and poorly understood conditions, with international assessments over time placing somewhat greater relative emphasis on the underlying design flaws than initial Soviet era investigations, which had focused primarily on operator error. A widely cited technical analysis of the RBMK design flaws specifically identified the combination of the positive void coefficient, the flawed graphite tipped control rod design, and instability at low power operating levels as together producing a reactor configuration capable, under a specific and admittedly narrow set of operating conditions, of undergoing precisely the kind of uncontrolled power surge that ultimately occurred, a conclusion that places Chernobyl squarely within this book's broader argument that catastrophic runaway events are, overwhelmingly, the product of identifiable, analyzable, and in principle preventable design and procedural gaps rather than of some fundamentally unpredictable natural phenomenon.

The disaster's aftermath produced significant, lasting design reform across the global nuclear industry, echoing the pattern of regulatory response following Seveso and, to a lesser degree, following West, Texas, examined in earlier chapters of this book. Remaining RBMK reactors still in operation, primarily within Russia, have since been retrofitted with a range of safety modifications specifically targeting the design flaws identified after Chernobyl, including redesigned control rods that eliminate the dangerous initial reactivity spike, additional neutron absorbers, and modifications intended to reduce the reactor's positive void coefficient under low power conditions, changes that safety analysts credit with allowing the remaining fleet of RBMK reactors to operate for decades since without a comparable incident. It is worth noting, in the interest of the evenhanded treatment this book has tried to maintain throughout its historical case studies, that the broader question of nuclear power's overall safety record relative to other major energy sources remains a subject of ongoing public debate and differing expert assessment, one this book does not attempt to resolve, since Chernobyl's inclusion here is specifically as an illustration of positive feedback runaway dynamics rather than as a basis for any broader judgment about nuclear energy policy.

What Chernobyl offers this book, above all, is confirmation that the fundamental pattern explored across its chemical and industrial chapters, a system capable of accelerating its own runaway once a specific threshold is crossed, absent an adequate and properly engineered mechanism to arrest that acceleration, is not confined to chemistry in the strict sense of molecular bonds breaking and forming. The same mathematical shape, a process whose rate of energy release increases as a function of its own progress, appears whether the underlying physics involves the oxidation of hay, the exothermic decomposition of ammonium nitrate, the electrochemical breakdown of a lithium battery's internal separator, or the neutron physics of a nuclear reactor's void coefficient. The specific triggering mechanisms differ enormously from case to case, and this book has taken considerable care to explain each one on its own terms rather than flattening the genuine scientific differences between them. But the underlying shape of the danger, a system whose own output feeds back to accelerate its own rate of energy release, recurs with a consistency that this book hopes, by this point, has become genuinely recognizable to its readers regardless of the specific scientific domain in which it next appears.

A second nuclear accident, occurring twenty five years after Chernobyl and involving an entirely different reactor design, a different triggering event, and, most relevant to this book, a distinctly chemical rather than purely nuclear physics mechanism, offers a useful complement to the Chernobyl case precisely because it demonstrates that catastrophic nuclear accidents are not always driven by reactor physics gone wrong. On March 11, 2011, a magnitude nine earthquake off the coast of Japan triggered a tsunami that struck the Fukushima Daiichi nuclear power plant, disabling the emergency diesel generators specifically responsible for powering the plant's cooling systems after the reactors had automatically shut down in response to the earthquake itself. Without power to circulate coolant, the nuclear fuel inside several of the plant's reactors, no longer undergoing active fission but still generating substantial heat from the ongoing radioactive decay of fission byproducts accumulated during normal operation, began heating uncontrollably, and as water levels within the reactor vessels dropped and the fuel rods became increasingly exposed to steam rather than liquid water, a chemical reaction distinct from anything examined in the Chernobyl case took over as the dominant hazard.

The fuel rods themselves were encased in a cladding made of a zirconium alloy, chosen for nuclear applications specifically because zirconium is comparatively transparent to the neutrons that sustain a reactor's fission chain reaction, an important design property under normal operating conditions that became catastrophically relevant for an entirely different reason once the fuel began overheating. At temperatures above approximately twelve hundred degrees Celsius, zirconium reacts with steam in a reaction that chemical engineers describe using the equation two water molecules plus zirconium yields two hydrogen molecules plus zirconium dioxide, a reaction that is strongly exothermic in its own right, meaning that once it begins, the heat it releases directly accelerates the fuel's own overheating, compounding the original loss of cooling with an additional, self reinforcing heat source, precisely the same runaway feedback structure this chapter has already examined in Chernobyl's positive void coefficient, here operating through ordinary, well characterized chemical oxidation rather than through neutron physics. Critically, this reaction also generates substantial quantities of hydrogen gas as a direct byproduct, and because the plant's ability to safely vent and recombine that hydrogen depended on electrically powered systems that had themselves been disabled by the same loss of power responsible for the cooling failure in the first place, hydrogen gas was able to accumulate within the reactor buildings rather than being safely managed and dispersed. On March 12, hydrogen that had built up on the upper floor of the Unit 1 reactor building found an ignition source and exploded, blowing the roof and upper wall cladding off the structure, and similar hydrogen explosions subsequently occurred at additional reactor units at the same site over the following days.

Estimates developed by subsequent engineering analysis suggest that the zirconium steam reaction at Fukushima's Unit 1 reactor alone generated on the order of several hundred kilograms of hydrogen gas, a quantity investigators concluded was more than sufficient to account for the observed force of the resulting explosion. The Fukushima case is instructive for this book precisely because its underlying chemistry, unlike Chernobyl's specialized reactor physics, is genuinely simple and entirely conventional: a metal reacting exothermically with steam at high temperature to produce a flammable gas is not a novel or exotic hazard, and the zirconium steam reaction had been characterized in nuclear engineering literature for decades before the Fukushima accident occurred. What made the reaction catastrophic at Fukushima was not any gap in scientific understanding but the specific and, in hindsight, foreseeable way that a single common cause, the tsunami's disabling of backup power, simultaneously defeated both the reactor's normal cooling system and the dedicated safety systems specifically installed to manage exactly this kind of hydrogen accumulation in an emergency, a design vulnerability in which a single external event undermined multiple, supposedly independent layers of protection at once, a pattern this book's readers will recognize as directly analogous to the failure of both the primary and backup overfill protection systems examined earlier in this book's account of the Buncefield explosion, where two nominally redundant safety systems turned out to share a common point of failure that neither had been designed to survive.

Reading Chernobyl and Fukushima together, separated by twenty five years, a different country, a fundamentally different reactor design, and a different immediate trigger, an operational testing error at one facility and a natural disaster at the other, reinforces this chapter's central argument with unusual force. Both disasters ultimately trace back to a runaway feedback process, positive void coefficient reactivity in one case, exothermic zirconium oxidation in the other, that a properly designed and properly maintained set of engineered safeguards should have been capable of arresting well before it produced the observed consequences. And in both cases, subsequent investigation identified specific, describable gaps, a control rod design that briefly increased reactivity at precisely the wrong moment, a hydrogen management system that depended on the same electrical power the initiating disaster had already destroyed, that transformed a serious but potentially survivable technical fault into one of the defining industrial catastrophes of its respective era. The chemistry and the physics, in both cases, behaved exactly as the well established science predicted they would. What failed, in both cases, was the engineering margin built around that well understood science, a margin that proved, when tested by the specific and admittedly severe conditions each disaster presented, to have been narrower than its designers had assumed.

It is worth closing this chapter by naming a distinction that separates Chernobyl and Fukushima from most of the other case studies examined throughout this book, since the distinction matters for how readers should weigh these two disasters within the book's broader argument. Nearly every other hazard this book has examined, hay fires, grain dust, aging peroxides, lithium battery cells, involves reactions whose worst case consequences, while genuinely severe and sometimes fatal on a considerable scale, remain fundamentally local and fundamentally bounded by the physical quantity of material actually present at the site of the failure. A nuclear reactor's stored energy, and correspondingly the geographic and temporal scale of a severe accident's consequences, radioactive contamination potentially affecting an area of hundreds or thousands of square kilometers and persisting, in some locations, for decades, operates on a fundamentally different scale, which is precisely why nuclear safety engineering has historically demanded, and generally received, a correspondingly higher standard of redundancy, containment, and regulatory oversight than most of the other industries examined in this book. That higher standard, evident in the extensive design reforms that followed both Chernobyl and Fukushima, reflects an appropriate and proportionate response to a genuinely higher consequence hazard, and it stands as a useful reminder that the practical lessons this book draws from its various case studies, while sharing a common underlying chemical and physical logic, are not meant to suggest that every exothermic hazard examined in these pages warrants an identical intensity of engineering and regulatory response. The scale of the potential consequence, as much as the underlying mechanism, must always shape how much caution a given hazard reasonably deserves.

Conclusion

The Fire That Was Always There

Return, for a moment, to the image that opened this book: a chemist named Robert, confident in a reactor that had run successfully one hundred seventy four times before, standing near a vessel that would kill him within minutes. That image was never meant to frighten. It was meant to demonstrate something considerably more useful than fear, which is the specific, recoverable gap between what a system appears to be doing and what it is actually doing, a gap this book has tried to close, chapter by chapter, across a genuinely wide range of settings that on their surface share almost nothing in common. A hay barn in rural Ontario, a fertilizer warehouse in central Texas, a battery inside a phone charging on a nightstand, a bottle of laboratory solvent forgotten at the back of a storage cabinet for a decade, a pan of oil left unattended for eighteen minutes on a stovetop. These scenes do not resemble each other in any obvious way. What connects them, and what this book has argued throughout, is that each one involves a chemical or physical process obeying the same handful of thermodynamic principles first formalized more than two centuries ago, principles that do not care whether they are operating inside a Texas fertilizer bin or a New York City apartment stairwell, and that produce, with total indifference to the setting, the same underlying pattern: heat generated faster than it can escape, accelerating itself until something gives way.

The specific stories collected in this book, T2 Laboratories, West Fertilizer, Bhopal, Seveso, Oppau, Texas City, the DeBruce grain elevator, the countless smaller and less publicized hay barn fires and laboratory peroxide accidents that never made national news but that killed and injured people all the same, were chosen not because they represent some unusually cursed corner of industrial history, but because each one offers a genuinely clear, well documented window into a specific exothermic mechanism, verified by careful investigation, government inquiry, or peer reviewed research, that a general reader can understand without a chemistry degree and can then recognize, transformed but structurally identical, in an entirely different context. The Coulomb explosion mechanism that makes sodium detonate in water, discovered through painstaking high speed photography in a Czech laboratory in the twenty first century, has no obvious connection to a battery failing after being dropped on a sidewalk, until the underlying principle, a reaction accelerating itself faster than any protective barrier can contain it, reveals the connection plainly. The bursting disc miscalibration that allowed the Seveso reactor to generate dangerous quantities of dioxin before finally venting has no obvious connection to a home cook reaching instinctively for water on a grease fire, until the shared lesson, that safety systems and safety instincts alike must be matched precisely to the actual chemistry of the hazard they are meant to address rather than to some generic, one size fits all notion of danger, becomes clear.

If this book has succeeded in what it set out to do, a reader finishing this final page should now notice things they would not have noticed before, not with anxiety but with the calm, specific attention of someone who finally understands what they are looking at. A stack of hay in a barn during its fourth week of storage.

A bottle of solvent in a garage or laboratory that has not been opened, or checked, or even glanced at in longer than anyone can quite remember. A phone battery that took a hard fall and still, apparently, works fine. A pot of oil on a stovetop, heating a little longer than intended while a phone call runs a few minutes over. None of these things are, on any given ordinary day, actually dangerous. The overwhelming majority of hay stacks, solvent bottles, dropped batteries, and unattended pans of oil complete their ordinary existence without ever becoming the subject of a chapter like the ones in this book. But each one is, in a specific and now hopefully recognizable sense, doing something, chemically or physically, whether or not anyone happens to be paying attention, and the entire argument of this book has been that paying the right kind of attention, informed, specific, and calm rather than anxious, general, and exhausting, is the single most effective and most underused tool available for making sure that doing something never quietly becomes doing something catastrophic.

The ice, to return one final time to this book's title, was never actually cold. It only appeared that way from a safe and unexamined distance, the distance most people maintain, entirely reasonably, from the specific chemistry of the objects and materials that fill their daily lives. This book has tried to close that distance, not to make the world feel more frightening, but to make it, in the very particular and useful sense this final chapter has tried to describe, more legible. Heat was there before the flame arrived, in every single story this book has told, and it will be there again, in situations and settings this book could never have anticipated or catalogued in full. The chemistry, as it has been since Lavoisier first measured it with an ice calorimeter in the 1780s, remains entirely indifferent to whether anyone is watching. Whether the flame eventually arrives, or whether the heat is noticed, understood, and addressed long before it does, remains, as it always has, a question this book hopes its readers are now considerably better equipped to answer.

A Note on Sources

This book draws on publicly available investigative reports, peer reviewed research, and established reference sources for its factual and historical content. Readers interested in exploring any of the incidents or scientific mechanisms discussed in greater depth are encouraged to consult the following categories of primary material.

Investigations conducted by the United States Chemical Safety and Hazard Investigation Board provide the most detailed available account of the T2 Laboratories explosion of 2007 and the West Fertilizer Company explosion of 2013, including detailed engineering analysis, timeline reconstruction, and formal safety recommendations. These reports are publicly available through the agency's own archives and remain the authoritative record of both incidents.

The Occupational Safety and Health Administration maintains historical data on grain elevator explosions in the United States, including the detailed investigative record of the DeBruce Grain Elevator explosion of 1998, and publishes ongoing guidance on the Grain Handling Facilities Standard first implemented in 1987.

The Health and Safety Executive of the United Kingdom maintains a detailed technical case study of the Seveso disaster of 1976, including analysis of the reactor design failures and control system inadequacies that contributed to the incident, material that informed this book's account of the event and its long term regulatory legacy in the form of the European Union's Seveso Directive.

The research describing the Coulomb explosion mechanism underlying alkali metal reactions with water was originally published in the journal Nature Chemistry by Philip E. Mason, Pavel Jungwirth, and collaborators at the Czech Academy of Sciences, and represents the most current scientific understanding of a reaction that has been demonstrated in classrooms for well over a century.

Historical accounts of the Oppau explosion of 1921 and the Texas City disaster of 1947 draw on multiple converging historical and scientific sources, including contemporary chemical industry analysis and subsequent engineering literature examining ammonium nitrate decomposition chemistry.

Contemporary data on lithium ion battery thermal runaway incidents, including figures on e bike and e scooter fires and aviation related incidents, draws on research compiled by fire safety research organizations and standards bodies actively engaged in ongoing battery safety monitoring, whose published figures reflect the most current data available at the time of writing.

The foundational thermochemical concepts discussed throughout this book, including the historical contributions of Antoine Lavoisier, Germain Hess, Marcellin Berthelot, and Svante Arrhenius, are drawn from the well established history of physical chemistry as it is documented in standard chemistry reference works and the history of science literature.

Data on residential cooking fires and cooking oil autoignition behavior draws on the National Fire Protection Association's published statistical reports and on peer reviewed fire protection engineering research, including work conducted at the University of Maryland's Department of Fire Protection Engineering.

The psychological framework discussed in the book's closing chapters, particularly the psychometric paradigm of risk perception, originates with the research of Paul Slovic and collaborators, whose foundational work established much of the modern scientific understanding of how the dread and unknown risk dimensions shape human judgment about hazard and danger. The related concept of the availability heuristic, discussed alongside the psychometric paradigm, originates with the research of Amos Tversky and Daniel Kahneman, whose work on judgment under uncertainty, developed beginning in the early 1970s, remains foundational to the modern study of cognitive bias in risk perception and decision making.

The account of the Piper Alpha disaster and its regulatory aftermath draws on the official findings of the Cullen Inquiry, published in 1990, along with subsequent technical analysis published by the Fire and Blast Information Group, an organization established directly in response to the disaster's findings. The account of the Texas City refinery explosion of 2005 draws on the United States Chemical Safety and Hazard Investigation Board's investigation and on the independent review commissioned by BP and chaired by former Secretary of State James Baker, generally referred to as the Baker Panel report.

The account of the Beirut port explosion of 2020 draws on multiple independent technical and investigative sources, including analysis published by Human Rights Watch, the research and advocacy organization Forensic Architecture, and peer reviewed engineering literature examining the blast's physical characteristics and underlying causes. The account of the Fukushima Daiichi accident of 2011 draws on technical analysis published by the World Nuclear Association and on peer reviewed engineering research examining the specific chemistry and quantity of hydrogen generated through the zirconium steam reaction during the accident.

The account of the Centralia mine fire draws on historical and scientific literature examining spontaneous coal combustion, including research on the specific thermal dynamics of underground coal seam fires and their documented history across multiple decades. The accounts of the MGM Grand Hotel fire of 1980 and the Kiss nightclub fire of 2013 draw on fire protection engineering analysis of flashover dynamics and on the historical record of subsequent fire code reforms adopted in response to each disaster. The account of the 2010 Texas Tech University laboratory accident draws on the United States Chemical Safety and Hazard Investigation Board's published case study of the incident.

Every effort has been made to present the factual content of this book accurately and to attribute historical and scientific claims to their appropriate documented sources. As with any work synthesizing material across a wide range of technical and historical subjects, readers with a specific professional or safety related interest in any topic covered here are encouraged to consult primary regulatory and scientific literature directly rather than relying on this book, or any single popular science work, as a substitute for professional guidance in matters of chemical safety, laboratory practice, or industrial process design.

www.ingramcontent.com/pod-product-compliance
Lightning Source LLC
LaVergne TN
LVHW100143061126
851857LV00010B/218
* 9 7 9 8 2 3 5 1 8 4 4 3 5 *